Questioning the Universe

Concepts in Physics

Questioning
the Universe
Concepts in Physics

Ahren Sadoff

CRC Press
Taylor & Francis Group
Boca Raton   London   New York

CRC Press is an imprint of the
Taylor & Francis Group, an **informa** business

A TAYLOR & FRANCIS BOOK

Chapman & Hall/CRC
Taylor & Francis Group
6000 Broken Sound Parkway NW, Suite 300
Boca Raton, FL 33487-2742

© 2009 by Taylor & Francis Group, LLC
Chapman & Hall/CRC is an imprint of Taylor & Francis Group, an Informa business

No claim to original U.S. Government works
Printed in the United States of America on acid-free paper
10 9 8 7 6 5 4 3 2 1

International Standard Book Number-13: 978-1-4200-8258-6 (Softcover)

Library of Congress Cataloging-in-Publication Data

Sadoff, Ahren.
   Questioning the universe : concepts in physics / Ahren Sadoff.
      p. cm.
   Includes bibliographical references and index.
   ISBN 978-1-4200-8258-6 (acid-free paper)
   1. Physics--Popular works. 2. Universe--Popular works. I. Title.

QC24.5.S33 2008
530--dc22   2008041364

Visit the Taylor & Francis Web site at
http://www.taylorandfrancis.com

and the CRC Press Web site at
http://www.crcpress.com

Contents

Chapter 13

Quantum Mechanics 155

Chapter 14

The Standard Model of Elementary Particle Physics 173

Preface

This book is aimed at you, the nonscience student. At the end of the course you are taking, I hope you will have a greater appreciation of what physics is, what physicists do, and how they do it.

I like to look at a course such as this as being analogous to a music appreciation course where students can learn a great deal about music without having to learn to read musical notes. And like music, I hope you will learn to appreciate the aesthetic aspects of science, and in particular physics. After all, what we will be dealing with is the most amazing, mysterious, and yes, most beautiful structure known to man— the universe itself. The spirit I hope to convey in this book is very nicely stated in a quote from Warren Weaver, former president of the American Physical Society:

> **Pure science is not technology, is not gadgetry, it is not some mysterious cult, it is not a great mechanical monster. Science is an adventure of the human spirit: it is an essentially artistic enterprise stimulated largely by disciplined imagination, and based largely on faith in the reasonableness, order and beauty of the universe of which man is a part.**

There are some words here that might surprise you and to which you might want to give some thought: "adventure of the human spirit," "artistic," "disciplined imagination," "faith," and "beauty." These are words not usually associated with science, but by the time you have completed this course, I hope you will agree with and appreciate Weaver's statement.

Part of the beauty of physics is its connectiveness or unity. I suspect those of you who have had some physics before might think just the opposite. Your impression was probably that physics was about a lot of unconnected topics like motion, forces, heat, sound, electricity, and so on. My hope is to convey to you the relationships among many of these subjects. We believe today that mother nature is quite efficient in that there are a relatively small number of laws that govern the universe. We may not know them all yet, but at the center of physics is the faith that nature is indeed understandable and that someday we will see her true beauty. She may try to hide it from us, but part of the fun for the physicist is trying to overcome her little tricks. The purpose of this book is to tell you the story of what we have found out about nature so far and how we have done it.

One important part of that story is the atom. It is the building block of all common materials with which we are familiar. In many of the topics we will be discussing throughout this book, the atom will be the key.

Scientists are curious people. Basically that means we ask questions (I suspect some of you thought of the other meaning of curious). In particular, physicists want to know about one or more aspects of the basic laws that govern the universe around us. That is why I have chosen the title I have for this book. By the time you are finished, I hope you too will be a little more curious about all sorts of things and will have learned to become better questioners.

One question many of you already have is: "How much math will I have to do?" If this were a book aimed at scientists or engineers, there would be a good deal of emphasis on math and problem-solving techniques. After all, that is what professional scientists and engineers have to do. But the emphasis here is going to be on learning *about* physics, not learning how to *do* physics. There may be a few arithmetic problems you will be asked to do, but they should be quite straightforward. Their purpose will be to help enhance your understanding of the concepts that you will be learning. In addition, there will be a good number of equations throughout the book. Equations are the best way physicists have found to express the laws of nature. They tell a story. So, we will want to look at these equations to be able to understand what they are telling us about nature. We will not, on the other hand, be doing very much at all as far as manipulating these equations to solve some mathematical problem.

Finally, throughout the book, there are going to be places formatted in shaded boxes where I suggest you think about something or try to answer some question before reading further. I strongly urge you to do this for your own benefit. The best way to learn is to be an active, engaged participant. It is well known that passively reading without some thought or introspection helps very little in your understanding or absorption of the material. On the other hand, if you force yourself to think about the problem, even though you may not come up with the solution, your understanding of the material will be greatly enhanced.

Acknowledgments

I would like to thank a number of people whose support and encouragement made this book possible.

I am very grateful to my editors at Taylor & Francis. John Navas, senior editor, initially read my manuscript, encouraged me to have it published, and has been very supportive throughout the entire process. In addition, he was extremely helpful in getting some of my Word-produced figures into publishable form. Amber Donley, project coordinator, and Judith Simon, project editor, guided me through the intricacies of going from the raw manuscript to a finished book. They made the process seem relatively easy. I can imagine it could have been much harder.

There have been literally thousands of students who have taken the course on which this book is based. I want to thank them for their attentiveness and feedback. The fact that the enrollment has stayed at a high level, allows me to assume that they have found the course valuable and have given their fellow students good reports. I hope they have left the course more scientifically literate and having a better idea of what physics is and what physicists do.

My wife, Barbara has given me continual encouragement to complete this project. In addition, for 50 years, she has had to put up with living with an ever-questioning physicist.

This book is based on a course that was developed at Ithaca College and Cornell University. I want to thank some of my former Ithaca College colleagues for their support in initiating this course. But the idea of a book started to become a reality while teaching at Cornell.

Don Holcomb was chair of the Cornell physics department in 1986 when I asked whether there would be any interest in my teaching a "physics for poets" course for one semester during my sabbatical leave from Ithaca College. I appreciate Don's willingness and support in giving me the opportunity to do so for that year.

The person who I am most indebted to and to whom I want to acknowledge my deepest appreciation is Doug Fitchen. Unfortunately, Doug tragically died earlier this year. Shortly after taking on the position of chair in 1987, after Don's term had ended, Doug called to ask whether I would consider teaching the course again the next year. I have now been teaching the course at Cornell for the past 22 years. Doug remained chair, on or off, for 15 of those years. He gave me both encouragement and support in more ways than I can enumerate here. I am forever grateful and appreciative for his interest and caring. In many ways, he is responsible for the fact that my ideas and words have ended up in this book. I only wish he were alive to see its publication.

Ahren Sadoff
Ithaca, NY

The Author

Ahren Sadoff received his BS in physics from the Massachusetts Institute of Technology in 1958. He was then awarded a Woodrow Wilson Fellowship for his first year of graduate study at Cornell University, where he received his PhD in elementary particle physics in 1964. The following year he served as a postdoctoral research associate at the Laboratory of Nuclear Studies (LNS) at Cornell.

In 1965 he joined the faculty of Ithaca College and remained there until 2000, when he retired as professor emeritus. He started the physics program, hiring all of the original faculty and developing a full physics curriculum. From 1966 to 1973, he served as chair of the physics department.

During the entire time he was at Ithaca College, and up to the present, he has been involved in elementary particle physics experiments at Cornell. He was an original member of the CLEO collaboration, which was comprised of as many as 250 physicists from about 20 academic institutions. This collaboration led the world for 20 years in the study of the b quark, one of the fundamental building blocks of matter. The collaboration has produced over 350 publications, primarily in b and c quark studies. He has presented numerous invited talks in such places as Rome, Naples, and Capri in Italy. He has also lectured in Spain, Singapore, Israel, Germany, England, and France, as well as in the United States.

In 1972, he was on sabbatical leave as a visiting scientist at the ADONE particle accelerator facility in Frascati, Italy, which is just outside of Rome. In 1980 he spent another sabbatical as a staff scientist in the Biomedical Division at the Lawrence Berkeley Laboratory in California. There he was involved in studies using beams of atomic nuclei for cancer treatment.

During a third sabbatical in 1986 at LNS, he started teaching a conceptual physics course at Cornell aimed at nonscience students. At that time, he was given an appointment as visiting professor. Upon his retirement from Ithaca College, he was appointed professor of physics at Cornell. He continues to teach this course, which attracts about 120 students a year. During this time, he also introduced a course titled "Concepts in Modern Physics" aimed at first-year physics majors.

Throughout his career, he has been very concerned about the public's understanding of science and the poor state of science literacy in the United States. He has been involved in many education and outreach projects. One such project was the production of a 30-minute video aimed at the public explaining the purpose and functioning of the particle accelerator facility at Cornell housed at Wilson Laboratory (renamed Laboratory for Elementary Particle Physics [LEPP]). More than 1,000 visitors a year view the video before taking a tour of the laboratory. It is also made available to any teacher to show in class. He also produced a brochure entitled "The Science at Wilson Laboratory," which describes the range of activities at the Laboratory, from elementary particle research to particle accelerator development to the many applications in x-ray science. In addition, Dr. Sadoff has written a manual for high school physics teachers explaining the standard model of elementary particle physics. This has become one of the required topics in the high school physics curriculum in New York State. All of these are available on the LEPP/Education Web site.

The laboratory has a very full and varied program in education and outreach, including programs aimed at elementary school students, teachers, and the general public. In 2002, a full-time education/outreach coordinator was appointed, and from the position's inception until August 2008, Professor Sadoff had been the supervisor.

Questioning
the Universe
Concepts in Physics

1 Units and Powers of 10

1.1 UNITS

How tall are you? You could answer 5 feet 8 inches, or 68 inches, or 1.73 meters. These are, of course, all the same heights, but the numbers are different because they are given in different units of length. Since science, in general, and physics, in particular, deal with measurement, in either an experiment or a prediction of an experimental outcome, we want to clearly define the units we will be using. Not only will this be important for clarity, but it can have very important practical consequences. Some of you may have heard that a scientific experiment failed several years ago because manufacturers of different parts of the apparatus used different systems of units while each one assumed the other was using the same units as they were.

We have found we can describe nature in terms of four fundamental quantities: length, mass, time, and electric charge. All other quantities can be expressed in terms of these four.

Length can be measured in the English system of units of inches, feet, and miles, or in the metric system of millimeters (mm), centimeters (cm), meters (m), and kilometers (km). We will use the metric system as most scientists normally do. Smaller or larger lengths can be expressed as multiples of powers of 10, as we will discuss below. For comparison, most of you probably know that there are 2.54 cm in 1 inch, and about 30 cm in 1 foot. A meter is a little over a yard.

We will also use the metric system for mass, which is the gram (g) or kilogram (kg). The prefix *kilo* means 1,000, so a kilogram is 1,000 g, just as a kilometer is 1,000 m. A kilogram has a weight of about 2.2 pounds. But be careful here, a kilogram is a unit of mass, but weight is a force. They are not the same thing, as we will discuss in Chapter 4.

Time, of course, is measured in seconds (s or sec), minutes (min), or hours (h). Most of the time we will use seconds.

Electric charge is measured in units of coulombs, named after Charles Coulomb (1736–1806), who devised the basic force law between two electric charges. One coulomb is a rather large amount of charge given that the charge on an electron or proton is 1.6×10^{-19} coulombs.

In this book, we will primarily use the MKS (meter-kilogram-second) system of units. This is the system most physicists use. As stated above, all other quantities can be expressed in terms of these three units. Listed below are some common quantities in terms of the basic units. When doing a unit, sometimes called a dimensional analysis, we will designate this by writing [L], [M], and [T].

Speed: [L]/[T]

Acceleration: [L]/[T]^2

Momentum: [M][L]/[T]

Force (Newtons): $[M][L]/[T]^2$

Energy (Joule): $[M][L]^2/[T]^2$

The words in parentheses are the names given to those quantities for the MKS system. In the English system, force is measured in pounds and energy in foot-pounds. To give some feeling for the relation between Newtons and pounds, 10 Newtons is very close to 2.2 pounds.

1.2   POWERS OF 10

What is the size of an atom? The answer depends somewhat on what particular atom is of interest. For the simplest atom, which is hydrogen, the answer is about 0.0000000005 m. What is the speed of light? Here, the answer is 300000000 m/s. In both cases, the way these numbers are expressed is very cumbersome. It is much more convenient and useful to use powers of 10, sometimes called scientific notation. Using this notation, the size of the hydrogen atom is 5.0×10^{-10} m and the speed of light is 3.0×10^8 m/s.

Let us review how we put a number into a power of 10. For the size of the hydrogen atom, note that we moved the decimal point ten places to the right to get it to be just after the 5. If that is all we did, we would have increased the number by 10 powers of 10. To correct for this, we divide by 10 powers of 10. But remember that $1/10^{10} = 10^{-10}$. Similarly, for the speed of light, we had to move the decimal point eight places to the left to get it just after the 3. If that is all we did, we would have incorrectly decreased the speed of light by 8 powers of 10. To correct for this, we multiply by 10^8.

In expressing numbers in scientific notation, we normally write the number multiplying the power of 10 as a number between 1.0 and 9.999—. So, while 55×10^2 is not incorrect, it is more conventional to write it as 5.5×10^3. This also helps us to know how accurate a particular number is (i.e., how many significant figures should be used). For instance, if a number is written in nonscientific notation as 35,500,000, this implies that it known to eight significant figures. But if it is written as 3.55×10^7, this clearly means it is accurate to three significant figures. If it were accurate to four figures, we would write it as 3.550×10^7. In what we will be doing throughout this book, usually two or three significant figures will be sufficient.

Since physicists are interested in the entire universe, from the smallest to the largest, we will be dealing with very large or very small numbers. At times, we will need to do some simple manipulations with these numbers, like multiplication or division. Let us remind ourselves how this is done. We will do it first generically, and then do a specific example.

For multiplication of two numbers, $a \times 10^m$ by $b \times 10^n$, we simply multiply a and b and add the powers of 10 to get $(ab) \times 10^{(n+m)}$.

$$(a \times 10^m) \times (b \times 10^n) = (ab) \times 10^{(n+m)}$$

To give a specific example: $(6 \times 10^4)*(3 \times 10^6) = 18 \times 10^{10} = 1.8 \times 10^{11}$. For one more example: $(6 \times 10^4)*(3 \times 10^{-6}) = 18 \times 10^{-2} = 1.8 \times 10^{-1}$, which could also be expressed as 0.18.

For division, we divide a by b and subtract the powers of 10.

$$(a \times 10^m)/(b \times 10^n) = (a/b) \times 10^{(m-n)}$$

Using the same numbers as above, we get $(6 \times 10^4)/(3 \times 10^6) = 2 \times 10^{-2}$ and $(6 \times 10^4)/(3 \times 10^{-6}) = 2 \times 10^{10}$.

QUESTIONS/PROBLEMS

1. How many meters in a yard?
2. What is the mass, in kilograms, of a 10-pound weight?
3. The speed of light is 186,000 miles/s. What is the speed of light expressed in meters/second?
4. A nanosecond is 10^{-9} seconds. What distance does light travel in 1 nanosecond? Give your answer in both centimeters and feet.
5. Write the following numbers in scientific notation:
   a. 1,000
   b. 1/10,000
   c. 2,500,000
   d. .000025
   e. 1
6. Calculate the following using and giving the answers in scientific notation:
   a. $.001 \times 10^4$
   b. $(5 \times 10^{-4})/(2 \times 10^{-3})$
   c. $10^6/.01$
   d. $(3 \times 10^{35}) \times (7.5 \times 10^{-3})$
   e. $(24 \times 10^{-5})/(.8 \times 10^2)$
7. A light year (LY) is defined as the distance light travels in the time of 1 year. Physicists use the symbol c to designate the speed of light. The value of c is

$$c = 3 \times 10^8 \text{ m/s} = 186,000 \text{ miles/s}$$

   a. Find the value of 1 LY in both meters and miles. Before you put in any numbers, first write the appropriate equation you need to get your answer.
   b. The sun is 93 million miles from earth. Find the time in minutes for light to travel from the sun to the earth. Again, first write the appropriate equation.

2 Physics and Its Methodology

2.1 WHAT IS PHYSICS?

A discipline can be defined by two criteria: the subject matter of study and the methodology by which the study is carried out. So, first let us talk about the subject matter; i.e., what is the definition of *physics*? Most definitions are basically a laundry list, like the one listed in the introduction of this book. Let us consider one that conveys the spirit of physics, although I imagine some might disagree with it. Anyway, here is my definition:

Physics is the search for the basic laws that govern the universe around us.

There are several assumptions that are intrinsic in this definition. Can you see what they are? Try to see if you can find them before reading further.

(By the way, we will be talking about assumptions several times in this book. It is especially important to be aware of them. Many times we take them for granted and do not even realize that we are making any assumptions, and so, of course, do not question whether or not they are true.)

OK, here is my list (maybe you can find even more):

1. We assume there are Laws to be found. (We will talk more about "Laws" a little later).
2. We also assume that we humans are capable of finding them.
3. We assume universality. This means that the laws we find here on earth are true everywhere else in the universe.

One other comment should be made about the definition. It has to do with the phrase "universe around us." That means from the very smallest objects that exist to the largest. Try to think what you believe are the smallest and largest objects that exist.

2.2 METHODOLOGY

So now that we have our definition, how do go about discovering these laws? In other words, what is the methodology of physics (which is actually the methodology of science, in general)? Basically we ask questions—hence the title of this book. These questions will hopefully lead us to answers or more questions. As an example, it might be useful to consider the following allegory.

2.2.1 THE FIRST SCIENTIST

Let us imagine the first scientist possibly being a caveman who noticed that the bright orb in the sky (we, of course, call it the sun) appears in the east and disappears in the west. He begins to notice that this happens every day (let us not worry about cloudy days). This is a prerequisite for the beginning of science—to notice correlations or relationships. If he were indeed the first scientist, his curiosity would lead him to question: Why does this keep occurring every day? This, in turn, would cause him to look for an explanation, i.e., a theory. His theory would probably be the most obvious one—that the sun goes around the earth traveling from east to west. This would certainly explain the rising and setting of the sun that he has observed. He, being the scientist he is, is quite excited by this realization and so naturally shares his idea with a friend.

The friend, also having a scientific bent, thinks about this and decides she has a competing theory. She proposes that the sun does not move at all, but the earth instead revolves about an axis spinning from west to east, so that it only appears to someone on the earth that the sun is moving from east to west. There would then, most likely, ensue a discussion (argument?) comparing the two ideas. One objection to the second theory (the heliocentric theory) would be that if the earth were spinning about an axis, why wouldn't everyone be thrown off? (This indeed was one of the arguments made against the heliocentric theory when it was proposed by Copernicus some 30,000 years later.) This objection would have most likely won the day, as it did for most people in the sixteenth century. Today, we would ask which of the two theories fits the known experimental facts and which predicts new observations. We will get back to this discussion soon.

2.2.2 WHY DO YOU BELIEVE?

Most likely you firmly believe that the heliocentric theory is the correct one, and that indeed the earth both revolves about the sun and spins around a north-south axis once every 24 hours. But it would be instructive to give some thought as to why you do believe this. Have you made your own observations or know of some experimental fact that proves this? And if not, why do you believe it? It might be interesting for you to see if you can understand your own belief system.

2.2.3 BACK TO THE QUESTIONS

Can we ask any questions we want? The answer is no. We are only allowed to ask certain kinds of questions—the kinds that will help us in our search. These type of questions are process questions, what we might call *how* questions as opposed to *why* questions. A *why* question implies a final cause or an intelligence. The philosophers have a fancy name for such questions. They call them *teleological* questions.

We have to be careful here. Every question that begins with *why* is not necessarily teleological; in fact, most are probably not. For instance, "Why is the sky blue?" is really a process question, which could be rephrased as "What is the process that causes the light from the sky to appear blue?" On the other hand, the question "Why was the universe created?" is certainly teleological since it implies that some intelligence decided one day to create the universe for some reason. But "How did the universe begin?" is a perfectly valid scientific question. In fact, the branch of physics concerned with this question is known as cosmology.

2.2.4 How Do We Answer the Questions?

We can answer this with one word: observation. A fancier way of saying this is that science is empirical. It starts and ends with experimental data. We saw that in our caveman story, where it was the observation of the sun rising and setting that led to a theory. And when there is more than one possible theory, which one is accepted will finally be based on the experimental observations. Even if there is only one proposed theory, it still must pass all the experimental tests. If it does not, then we know that that theory, no matter how well it seemed to work, cannot be entirely correct.

We have used the word *theory* several times already without really defining it, even though most of you probably have a reasonable idea of what the word means. We will have a fuller discussion of exactly what a theory is below, but first it is important to discuss the uniqueness of the experimental method in the natural sciences in contrast with the social sciences. Experimenters in the natural sciences have the unique ability to control a relatively small number of variables and change only one at a time, knowing that all the other variables remain constant and unchanged. This ability to control the variables and be able to change only one at a time is the key to the experimental method. If one performs the same experiment as many times as he or she likes, one expects to get the same results. In the social sciences, this is not necessarily true. Here we are dealing with thinking and feeling human beings, each of whom could very well react differently to the same situation. In fact, the same individuals could react differently at different times. If there are any social scientists reading this, some of them might take exception to the above. But good social scientists understand this point very well. So, let us repeat this very important statement: the great success of the scientific method in the natural sciences is in the ability to be able to control the variables and know with reliability that it was done correctly.

This does not imply that errors cannot be made; of course, errors have and will be made. Scientists are human beings and, as such, are certainly not infallible. But finally, independent analysis should discover any mistakes. In fact, just about every working scientist is a professional skeptic. We are always asking when presented with a new result, "OK, where's the goof?" This is especially true of our own work. After all, we would much prefer to find our mistakes before we make our results public, instead of someone else finding them.

2.2.5 THE NEED TO BE QUANTITATIVE

Experiments, by their nature, are quantitative. A measurement results in a number, whether in units of time, length, mass, etc. This, then, requires us to frame our theories quantitatively, and thus express predictions in terms of measurable quantities. For example, the often asked question "Why is the sky blue?" is not really a very good scientific question. Blue is not a measurable quantity. After all, what does *blue* mean to someone who is color-blind? A correct scientifically expressed question is: What causes the light from the sky to have maximum intensity at short wavelength? For now, let us ignore the fact that this is a rather pompous way of talking; the point is that the quantity wavelength is measurable. So, at times, we will have to be careful about our use of language to ensure that what we are talking about is unambiguous.

This quantitative aspect also imposes the use of mathematics as an important part of the language of physics. Mathematics is not a science. It fails that definition as to both its subject matter and its methodology. But it is a very useful tool, and every working physicist has to be able to use it, some more, some less. The theories we have developed are expressed as mathematical equations. We have found that this is the most efficient way of expressing nature's laws. In addition, this allows us to efficiently combine different theories to obtain new results. As said earlier in this book, we will be doing relatively little mathematical manipulations of equations, but we will be considering a good number of equations to help us understand what they are telling us about the nature of nature.

2.2.6 THEORIES

We used the word *theory* earlier since you probably already have some reasonable idea what it means. But now we want to discuss theories in some detail. You may be surprised at some things you learn here.

> What exactly is a theory? Before you read further, can you think of some other words that could be synonyms?

Basically, a theory is a guess. Usually an educated guess, but a guess nevertheless. Other words could be *hypothesis*, *conjecture*, or *idea*, but the word *guess* makes the point. This guess must relate certain phenomena to each other and explain the relationship. For instance, our caveman's theory related the rising and setting of the sun every day and explained it by hypothesizing that it was due to the sun's motion circling the earth. But remember, our caveman had a friend who came up with a different theory about the earth spinning on its axis. Both seemed to explain the same observations. So, how do we choose?

We choose by subjecting each to the experimental test. If they are indeed different theories, they must make some predictions that are different—not necessarily all, but some. So, a very important criterion for any theory is that it must be testable; i.e., it must make new predictions that can be tested. A philosopher of science, Karl Popper, expressed this in an interesting way. He said that every theory must be *falsifiable*. That means it must have the ability to be falsified. It is another way of saying that a theory must make new predictions that can be tested, and if it fails the test, then it cannot be correct (i.e., it is false). Popper had an interesting way of looking at science that is worth considering here. Below are some quotes about his ideas from an April 1, 2002, *New Yorker* magazine article by Adam Gopnik.

> Science didn't proceed through observations confirmed by verification; it proceeded through wild, overarching conjectures, which generalized "beyond the data" but were always controlled and sharpened by falsification, by refutation, by the single decisive experiment. It was the conscious, purposeful search for falsifications, and the survival of the theories in the face of them, that allowed science to proceed and objective knowledge to grow.

And along the same line:

> Science wasn't a form of proof. It was a style of quarreling. The reason science gave you sure knowledge you could count on was that it wasn't sure you couldn't count on it. Science wasn't the name for knowledge that had been proved true; it was the name for guesses that could be proved false.

One consequence of this is that no theory can ever be *proven* in science. This is because no matter how many times a theory has been successfully tested, it could fail some new test tomorrow. Of course, after a theory has consistently passed all experimental tests, the scientific community will accept it as most likely being correct. But we still understand that it cannot be proven.

If that is the case, then what is the difference between what is called a theory and what is called a law? The answer is that there is no difference except for possibly some historical precedence. For instance, we know Newton's *law* of gravity is not quite correct (we need general relativity to better describe the gravitational force). On the other hand, we believe Einstein's special *theory* of relativity is most likely correct since, as of this writing, it has passed all experimental tests. Someone has suggested, somewhat cynically, that theories proposed before 1850 are called laws and those proposed after that date are theories. On the other hand, we have indeed used the word *law* in our definition of physics to describe the actual law of nature, which we are trying to discover.

2.2.7 Models

A model can be very useful in developing a theory. In some cases, the model is an intrinsic part of the theory; in other cases, the theory requires no model at all. Probably one of the most common models with which you may already be familiar is the planetary model of the atom. Listed below are three different types of models, and they all help us get a better picture of the physical system we are trying to understand.

1. The analogy: This is probably the most common type. It is a picture we have in our mind to help us a visualize something we cannot directly perceive. The planetary atom is just such a model. Here we think of the atom resembling our solar system, with the nucleus like the sun and the electrons rotating in their orbits like the planets. We will discuss the history of the different atomic models in more detail in Chapter 9.
2. The systemic model: This allows us to explain properties of a specific system in terms of the properties of a more generalized system. For example, is light (a specific system) a wave or particle phenomenon (generalized systems)? Since waves and particles have properties that are quite different, identifying light with one of them implies certain properties for light.
3. The mathematical model (do not worry, we will not do much with this): In doing the mathematics associated with some theory, we could have an equation that is recognized as the same mathematical equation associated with an entirely different part of physics. This could help give us insights into the new theory we are trying to understand. To give one example, the equation that describes the motion of a mass attached to a spring is exactly the same equation as the one that describes the oscillation of the electrical current in your radio or television tuner.

2.2.8 Aesthetic Judgments

From a formal point of view, a theory is accepted as long as it passes the experimental tests to which it has been subjected. But in practice, theories are subjected to unscientific aesthetic tests as well. Such terms as *simplicity, elegance,* and *beauty* have been used to describe or select one theory over another.

Physicist and historian of science Gerald Holton has written about this in his book *Thematic Origins of Scientific Thought.* What he means by *themata* are prejudices or presuppositions that every scientist has about the nature of nature. One example of such a prejudice is Einstein's famous statement "God does not play dice with the universe." This reflects Einstein's view as to how nature should behave in the case of the probabilistic nature of quantum mechanics even though there was no empirical evidence to support the implications of his remark. In fact, today the evidence strongly supports the interpretation that troubled Einstein.

Two other examples are taken directly from Holton's book. They involve judgments of two Nobel Prize-winning physicists about each others theories. In 1926, Heisenberg wrote: "The more I ponder the physical part of Schrodinger's theory, the more disgusting it appears to me." At about the same time, Schrodinger, in his turn, wrote about Heisenberg's approach: "I was frightened away by it, if not repelled."

The ironic part of the above quotes is that it was later shown that these were actually identical theories, just expressed in different mathematical forms. Certainly words like *disgusting* and *repelled* are not scientific, but such aesthetic judgments are used by scientists all the time.

2.3   END-OF-CHAPTER GUIDE TO KEY IDEAS

- What is the definition of physics, and what assumptions are associated with that definition?
- What is the basic methodology of science?
- What types of questions do physicists ask?
- What is a theory?
- What is a model?
- How do aesthetic judgments enter into physics?

QUESTIONS/PROBLEMS

1. What do you think are the smallest objects that exist? Answer the same question for the largest objects that exist.
2. Formulate at least three questions you have about the physical (as distinct from biological) universe around you.
3. What is the basic methodology of science?
4. List three scientific questions that could begin with *why*.
5. Is the question "Why is the sky blue?" a good scientific question? Explain your answer.
6. What is the important characteristic of experiments in the natural sciences?
7. Which of the following can be considered quantitative scientific terms: (a) short, (b) hot, (c) temperature, (d) green, (e) frequency?
8. Define *theory*.
9. Define *law*.
10. Define *model*.
11. A criticism of the theory of evolution is that it is only a theory and has not been proven. Discuss the validity of this criticism.
12. What does Karl Popper mean when he says that a theory must be falsifiable?
13. What can you say about a theory that has been falsified?
14. List three scientific models.
15. Is Newton's law of gravity correct? Discuss your answer.
16. Is Einstein's theory of relativity correct? Discuss your answer.
17. Should there be a place for aesthetic judgments in science? Discuss your answer.

3 Motion

Why start off with a discussion of motion as our first physics topic? There are several reasons. Certainly motion is very familiar to us. After all, we experience it every day. Historically, it was one of the first things studied by the Greeks. To them, there was the perfect and eternal motion of heavenly bodies, where the gods resided, and the imperfect motion of earthly objects that eventually had to come to rest. In our more modern world, we realize that just about everything is in some state of motion. The earth both revolves about the sun once a year and rotates on its axis once a day. The air around us is always moving, sometimes in an organized pattern that we call wind, but even on a perfectly calm, windless day, the air molecules are in random motion, causing the external pressure on us. Also, the whole idea of being in motion is a relative one, as both Galileo and Einstein have shown us. We will have a full discussion of this in the later chapters on relativity theory.

Finally, the most important reason to start with ideas of motion is that it is one of the main windows that allows us to understand the nature of forces. We can view the universe as being made up entirely of just four constructs: matter, forces, space, and time. Without matter, of course, there would be nothing in the universe. But without forces, the matter would not be able to interact, and thus would not be able to "clump" up to form atoms or molecules or cells or us. Before we can understand the nature of forces and how we can measure them, we have to first understand motion and how forces affect motion. Objects in motion move through space and time. We assume these to be absolute in the sense that the distance between two points in space will be the same for all observers, and the time interval for a given event will be the same for all observers. While this seems to be an obvious assumption, we will learn in the chapter on relativity that is it an incorrect one. For now, we can safely avoid these complications in our discussion of motion.

The word *motion* is really a very fuzzy word. Yes, we usually know when we are moving, but to understand exactly what that means (i.e., to be scientific), we have to carefully define the quantities (*variables* is a better word) that describe the nature of the motion. You should try to do this for yourself before reading further.

Here are the variables we will need to precisely define motion (hopefully you have come up with these yourself): position, distance, speed, velocity, acceleration, and time. There are several words here you might think mean the same thing, for instance, *position* and *distance*. But *position* is a point in space, while *distance* is the difference between two positions. In other words, if there are two positions, let us call them x_1 and x_2, then the distance, d, between these two positions is

$$d = (x_2 - x_1) \tag{3.1}$$

Also, *velocity* and *speed* are sometimes used interchangeably, but they have different meanings. *Speed* is a measure of only how fast you are moving, while *velocity* has two numbers associated with it: the speed and the direction of motion. Such quantities that require two numbers (a magnitude and a direction) to describe them are called vectors, while a quantity like speed, which is a magnitude, is called a scalar. The speed is the magnitude of the velocity vector. In some cases, we might only be interested in the magnitude of a vector quantity and can ignore directions. But most of the time, the full vector nature of a quantity will be important. In such cases, we will write the symbol with **bold type**. So, the velocity vector will be written as **v**, while the speed will be written as v. Vectors can be drawn as arrows, where the length of the arrow represents the magnitude of the vector and the direction the arrow is pointing indicates the direction of the vector.

3.1   RELATING THE VARIABLES OF MOTION

We have already defined distance as the difference between two positions. Speed is a measure of how fast your position is changing with time, i.e., the distance traveled in a time interval:

$$v = (x_2 - x_1)/(t_2 - t_1) \tag{3.2}$$

where x_2 is the position at time t_2 and x_1 is the position at time t_1. This can also be written using the symbol Δ for change or difference:

$$v = \Delta x/\Delta t \tag{3.3}$$

where $\Delta x = x_2 - x_1$ and $\Delta t = t_2 - t_1$.

Note that the above equation is actually an old friend that you all know. If we write d for Δx and T for Δt, then we have $v = d/T$, or $d = vT$. You use this equation many times without even thinking about it. How many times, when on a trip, have you tried to estimate how much time it will take you to reach your destination, knowing how far you have to go and what your speed is? Well, not only have you used this equation, but you have inverted it to $T = d/v$, probably in your head, to get the time.

Technically, v is the average speed over the time interval $(t_2 - t_1)$. The speed could be constant over this interval or changing throughout the interval; in either case, the average speed, as defined above, is still the same as shown below by the two plots of x versus t (Figure 3.1).

We have one more variable to define: acceleration. Acceleration is a measure of how fast the velocity is changing. Since velocity is related to a change in position, acceleration is a measure of a change of a change. This makes it harder to have a good intuitive grasp as to its meaning.

Formally, acceleration is defined as

$$\mathbf{a} = (\mathbf{v}_2 - \mathbf{v}_1)/(t_2 - t_1) = \Delta\mathbf{v}/\Delta t \tag{3.4}$$

Note that this is a vector. If we are only dealing with linear motion (one-dimensional motion), the acceleration only depends on the change of speed. But if the motion is

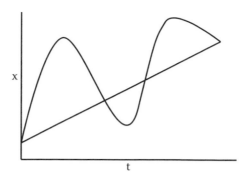

FIGURE 3.1 Graph of position as a function of time for two different types of motion. In both cases, the average speed is the same. For the straight line, the speed at any instant (the instantaneous speed) is the same as the average speed. For the curved plot, the instantaneous speed changes from instant to instant, but the average speed is the same as that for the straight-line motion since the total change in position is the same for the same time total interval.

taking place in two or three dimensions, then we have to take into account changes in direction as well as changes in speed. We will discuss two-dimensional motion below, but for now let us just consider one-dimensional motion for simplicity. This will allow us to understand most of what we need to know.

3.2 GRAPHS OF ONE-DIMENSIONAL MOTION

Those of you who have had a previous physics course probably remember that there were four or five different equations you had to know in order to solve motion problems. Except for the definitions above, we will not be using any of those equations. Instead, we will use graphs to help us understand how the variables we have defined relate to different types of motion.

Since we will be using graphs not only here but many more times throughout this book, let us remember some important things about them. Graphs are like pictures, and thus tell a story if we know how to read them. They tell us how one variable is related, or depends, on another variable. In order to read them, we have to know what is being graphed (i.e., what are the variables being related). In other words, the axes must be labeled so we know what the variables are. A graph without a label is meaningless. Also, since a graph shows how one variable depends on the other, by convention, the dependent variable is plotted on the vertical axis and the independent variable is plotted on the horizontal axis.

3.2.1 CONSTANT SPEED

There are four graphs below, all representing an object with constant speed. Before reading the text below the graphs, look at them and see if you can figure out what the object is doing. Note that the independent variable is time in all cases. We are usually interested in how an object's motion is changing with time.

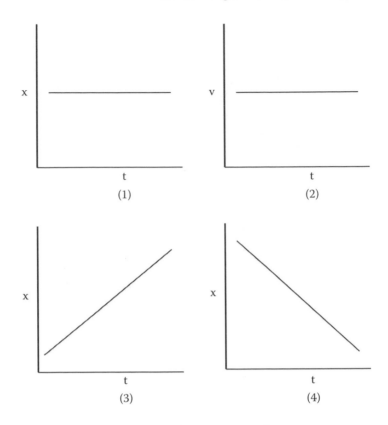

FIGURE 3.2 Graphs depicting motion with constant speed.

OK, let us see how you did.

> Graph 1: We see the position, which is plotted on the vertical axis (the depen-
> dent variable), is not changing. The object is not moving, so it has zero
> speed.
> Graph 2: This looks the same as graph 1, but the dependent variable is speed
> and it is not changing. This represents an object with constant positive
> speed. We define positive and negative speeds to distinguish between going
> in opposite directions. We can, for instance, call objects moving to the right
> positive, and those going to the left as moving in the negative direction.
> What we designate as positive or negative is arbitrary as long as we define
> one direction as positive (then the other has to be negative.) If we wanted
> to graph a constant negative speed, we would draw a horizontal line below
> the time axis.
> Graph 3: Here the position is the dependent variable, but it is increasing with
> time. So, the object is moving with a constant speed. How do we know the
> speed is constant? Let us consider the slope of the graph. Remember the
> slope of a line is rise/run, where the rise is the change in the y variable and
> the run is the change in the x variable. That is,

$$\text{rise/run} = \Delta x/\Delta t$$

So, we see the slope of the position versus time graph is, by our definition above, the speed. In this case, the slope is constant (since it is a straight line) and positive. Note that while this graph and graph 2 could depict the same motion, this graph has more information in it. You can obtain the speed from the slope, but you also know the position at any time. Graph 2 only tells you the speed. You can tell nothing about the object's position from this graph.

Graph 4: Since the line has a constant negative slope, the object has a constant negative speed.

3.2.2 CONSTANT ACCELERATION

Again, below are four graphs depicting the motion of an object with constant acceleration. As before, try to see if you can figure out the motion of the object. In this case, there are both positive and negative values included in the graphs.

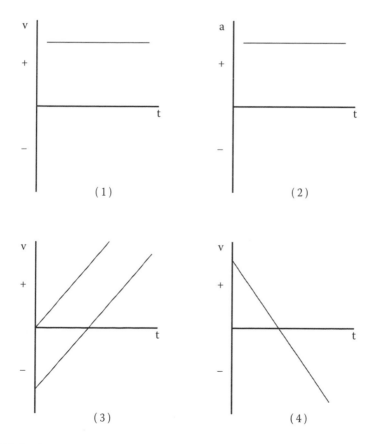

FIGURE 3.3 Graphs depicting motion with constant acceleration.

Graph 1: This is the same as graph 2 for constant speed. Since the speed is constant, the acceleration is zero. Even though zero is a constant, zero acceleration is a special case and we want to distinguish it from non-zero acceleration. Also, the line is drawn above the origin, which shows the object has a positive speed (e.g., it is moving toward the right). The object could have been moving to the left (negative speed), so the line would have been drawn below the origin. In either case, the acceleration is zero.

Graph 2: Here the acceleration is the dependent variable and has a constant positive value. If the line were drawn below the origin, the acceleration would be negative. We will see what is meant by positive and negative acceleration in discussing the next two graphs. It is important to note here, as we will below also, that negative acceleration does not necessarily mean deceleration.

Graph 3: Here there are two parallel lines, both indicating the same acceleration. How do we know that? As we did in graph 3 for constant speed, let us look at the slope and again calculate what it is:

$$\text{rise/run} = \Delta v / \Delta t$$

By our definition earlier, $a = \Delta v / \Delta t$. So, the slope of the graph of v versus t is the acceleration. Since the two lines are parallel, they have the same slope and hence the same acceleration. Then what is the difference between the two lines? The top line is always above the origin, so it always has a positive speed (i.e., it is moving in only one direction). On the other hand, the lower line starts off with a negative speed, which means, let us say, it is moving to the left. While its speed is negative, it is decreasing until it crosses the time axis.

At that point, the speed is zero. After that, the speed increases in the positive direction. Why do we say the acceleration is positive if the speed is decreasing? It is because *positive* or *negative* implies a direction, not whether the object is speeding up or slowing down. In fact, usually in physics, we do not use the word *deceleration* since acceleration can cause either an increase or decrease in speed, as we have seen in this example. We will see this further in the discussion of graph 4.

Graph 4: Here, we see that the line has a negative slope, so the acceleration is negative. As one example, this graph could depict the motion of a ball thrown straight upward. Let us analyze it carefully. Initially the graph shows that the speed is positive and has its largest value. If we define the positive direction as upward, then this is just what happens when we throw a ball upward. It leaves our hand with its highest speed, and then the speed decreases as it rises. At the highest point, the speed must be zero (where it crosses the time axis on the graph), and from there on it falls with increasing downward (negative) speed. Note that throughout the entire motion, the acceleration is always negative and constant—while the ball is moving upward, when it stops, and when it is moving down. We see this must be true since the line depicting the motion has a constant negative slope. Some of you may be having a hard time accepting this. It is most likely because you are confusing a quantity (the speed) with a change in that quantity (the acceleration). To try to understand this better, let us look at the graph just where the line crosses the time axis. Here the speed is zero at that instant, but the *change in speed* is not zero. The change is proportional to the slope, which is certainly not zero at that point or anywhere else on the graph. When we talk about forces in the next chapter, we will see another reason the acceleration of an object while in the air must be constant throughout the motion, including at the top, where the speed is zero.

3.3 TWO-DIMENSIONAL MOTION

In what we have just discussed above, we had to worry a little bit about the direction to the extent that an object could have a speed in opposite directions, such as left and right or up and down. But once we consider more than one-dimensional motion, the direction can be arbitrarily changing. In fact, this is the world we live in and experience every day as we are walking or riding in a car. We have a pretty good intuitive feeling about our velocity changing since we turn corners or move along some sort of curved path. Notice, I said *velocity*, not *speed*. In one dimension, if your direction of motion changes, your speed must first change to allow you to come to a stop before you can change your direction. We saw this in graphs 3 and 4 above. But in two dimensions, you could, for instance, go around a circle at constant speed.

In this case, your speed is constant but your velocity is continually changing. In fact, we will learn just about all we need to know by considering this single case of circular motion in two dimensions. Considering three-dimensional motion just complicates things unnecessarily.

The diagram below could represent the motion of a car going around a circle at constant speed, but *not* constant velocity. The arrows represent the velocity vectors at four different parts of the circle. The length of each arrow is the same, indicating constant speed, but the direction of each arrow is different, so the velocity vectors are not the same. From our definition of acceleration above, $\mathbf{a} = \Delta\mathbf{v}/\Delta t$, since the velocity vector is changing, this is accelerated motion even if the speed is not changing.

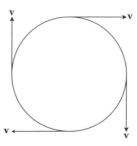

FIGURE 3.4 Velocity vectors for an object moving in a circle.

Let us discuss both the direction and the magnitude of the acceleration in this case of circular motion. Neither is obvious from our definition of **a** as given above.

The direction of the acceleration vector is inward toward the center of the circle at any point along the circle. To understand this qualitatively, consider the arrow at the top of the circle in Figure 3.4. If a car were at the point at the tail of the arrow, where it just touches the circle, and if it were going straight instead of following the curve, then in a short time the car would be at the tip of the arrow. But because the car is actually following the curve of the circle, it has "fallen" in closer to the center of the circle, compared to where it would have been if it were going straight. So we say that the car is accelerating inward. A rigorous mathematical treatment shows the $\Delta \mathbf{v}$ vector indeed points inward. This type of acceleration is known as *centripetal* acceleration since the direction of the vector points inward toward the center of the circle. A *centrifugal* acceleration would point outward, away from the center.

The magnitude of the acceleration is given by the expression $a = v^2/R$, where v is the speed and R is the radius of the circle. This looks nothing like $\Delta v/\Delta t$, but let us look at Figure 3.4 again. Consider the two velocity vectors at the top and bottom of the circle. (Note: This will not give us an exact answer, but it will allow us to see why the expression for the acceleration has the two powers of v and is divided by the radius.)

Remember that $\Delta \mathbf{v} = \mathbf{v}_2 - \mathbf{v}_1$, where position 1 is at the bottom of the circle and position 2 is at the top. Let us call the direction to the right at the top of the circle the positive (+) direction (we are free to call either direction positive, but then, of course, the opposite direction is negative). Then $\mathbf{v}_2 - \mathbf{v}_1 = \mathbf{v} - (-\mathbf{v}) = 2\mathbf{v}$. As for Δt, it is just the time interval for the car with a speed v to go from point 1 to point 2, or halfway around the circle. So we have to find the time to go a distance of half the circumference of a circle with radius R. Using our old friend $T = d/v$, we get $\Delta t = \pi R/v$ (remember the circumference of a circle is $2\pi R$). Thus,

$$\Delta v/\Delta t = 2v/(\pi R/v) = 2v*v/(\pi R) = (2/\pi)(v^2/R)$$

As stated above, this did not give us an exact answer, but hopefully we can see how the expression (v^2/R) comes about. One power of v comes from the change in the velocity vector, and the other power of v comes from the fact that the faster the car is moving, the smaller is the time interval to get from one part of the circle to another.

For those of you who are interested, let me try to explain why the argument above did not give the exact correct answer. What we did was to calculate the average acceleration over a half-circle. But since the velocity vector is changing at each instant (the direction is always changing), we have to calculate the instantaneous acceleration at each point in the motion. In order to do this, we have to either use calculus or consider the change of velocity during a very small part of the motion, not over such a large part of the circle.

At this point we have finished our discussion of motion. We are ready to go on to consider our next topic: Forces.

3.4   END-OF-CHAPTER GUIDE TO KEY IDEAS

- Why begin our first discussion of physics with ideas about motion?
- What are the variables needed to describe motion?
- How are these variables related?
- What is the difference between a scalar and a vector quantity?
- Can you read the different graphs of motion and tell what type of motion each graph describes? In other words, if each graph described the motion of a car, what does the graph tell you about the position, speed, and acceleration of the car?
- Why is circular motion accelerated motion?
- In circular motion, what determines the acceleration, and which way does the acceleration vector point?
- Why does the centripetal acceleration depend on v^2?

QUESTIONS/PROBLEMS

1. Why have we begun our discussion of physics with the idea of motion?
2. How are position and distance related?
3. What is the defining equation that our old friend $d = vT$ is based on?
4. Define *vector*. Give at least three examples of vector quantities.
5. What is the relation between speed and velocity?
6. Define, both in your own words and by an equation: *speed* and *acceleration*.
7. If A has a greater speed than B, does that mean that A has a greater acceleration? Justify your answer in writing and also by a graph.
8. Draw three different graphs (different vertical axes) depicting the motion of an object with (a) positive constant acceleration and (b) negative constant acceleration.
9. Draw a graph for:
   a. The position of an object going away from you with constant speed
   b. The position of an object coming toward you with constant speed
   c. The position of an object that is going away from you with constant, positive acceleration
   d. The speed of an object that is going away from you with constant, positive acceleration
   e. The speed of an object dropped from a building
   f. The acceleration of an object dropped from a building
   g. The acceleration of an object thrown upward from the ground
10. Each of the graphs below represents the motion of an object moving along a straight line. The first row are all graphs of position x versus time. The second row is of speed versus time, and the third row is of acceleration versus time. Also note that the x axis intercept is zero on the y axis.
    a. Give the letters for the graphs that represent nonzero constant acceleration.
    b. Give the letters for the graphs that represent nonzero constant speed.

c. Which graphs could represent the same motion? Of the ones you have chosen, which gives the most information about the motion?

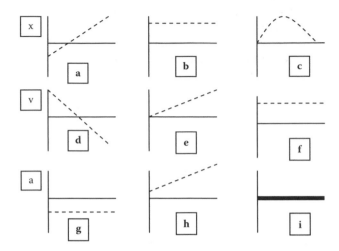

11. Can the velocity of an object be zero at the same time that its acceleration is not zero? Explain and give an example.
12. Can you assume a car is not accelerating if its speedometer shows a steady reading of 40 mph? Explain.
13. Does the odometer of a car measure a vector or scalar quantity? What about the speedometer?
14. If you go around a curve at constant speed, is this accelerated or nonaccelerated motion? If accelerated, what is the direction of the acceleration vector?
15. Will the acceleration be the same when a car rounds a sharp curve at 50 mph as it is if it rounds a gentle curve at the same speed? Explain.
16. What is the average speed of a sprinter who runs the 100-yard dash in 9.8 s? What would be the time for 1,500 m at this pace?
17. At an average speed of 11.8 km/h, how far will a car travel in 175 min?
18. Two cars go around the same curve. Car A has a speed of 30 mph, while car B has a speed of 60 mph. Which has the greater acceleration, and what is the ratio of their accelerations?
19. A boat moves in still water with a speed of 10 mph.
    a. How much time does it take to go a distance of 4 miles?
    b. If the boat is in a river that is flowing with a speed of 4 mph, how much time does it take to go 2 miles if it is heading downstream (in the direction the river is flowing)?
    c. How much time does it take to go 2 miles if it is heading upstream?
    d. What is the time to make the total down/upstream trip of 4 miles?
    e. Given your answer in part a, should this surprise you? Discuss.

4 Forces

4.1 THE FUNDAMENTAL FORCES

What is a force? One answer is that it is a push or a pull. A better answer, that we will find to be more useful, is that it is an interaction between two or more objects. For most of our discussion, two objects will suffice.

Forces are no strangers to us since we interact with all sorts of things every day. Below is a list of forces I have compiled. Before reading my list, it would be instructive for you to take out a piece of paper and make your own list. Hopefully you will come up with some not on my list.

Gravity	Electric	Weak nuclear	Strong nuclear	Centrifugal
	Magnetic			Centripetal
	Friction			
	Wind force			
	Contact force (between surfaces)			
	Muscular force			
	Chemical			
	Atomic			

I am sure you have noticed that my list is arranged in columns or categories. Let us look at the last column first. Both items are, in fact, not forces at all, but adjectives describing the action of a particular force. A centrifugal force is any force that is directed outward from the center of a curve when an object is traveling in curved motion. Similarly, a centripetal force acts inward toward the center of the curve.

Gravity is usually the force most people list first, as I have. It, of course, is very important to us since it keeps us bound to the earth and the earth to the sun. The second column contains many familiar forces under one heading. Why? Because all these seemingly different forces are all due to only one force. Electric and magnetic are not separate forces, but just different manifestations of what is known as the electromagnetic force (we will discuss this in more detail shortly). The force that holds the atom together is not some special new force, but is just due to the electrical attraction of the negatively charged electrons to the positively charged protons in the nucleus. Similarly, different atoms interact by the attraction or repulsion of the electrons and protons in one atom acting on the electrons and protons of another atom. All the other forces in the second column are due to the interaction of atoms (e.g., muscle atoms, air atoms in the wind), so all forces in that column are due to the electromagnetic force.

Finally, there are the strong and weak nuclear forces. They do not directly affect our daily lives since they only act over nuclear or subnuclear dimensions, which are much, much smaller than atomic dimensions. They do affect us indirectly since they are responsible for holding the nucleus together (so we can have atoms) and for certain types of radioactive nuclear decays. They are also very important in powering the sun, which is the ultimate source of energy for us here on earth.

So, of all the myriad forces we may know of, there are only four. We believe these four—gravity, electromagnetic, strong nuclear, and weak nuclear—are the fundamental forces of nature. But our most modern theories carry this reduction (or unification) of forces even further. We believe, but have not confirmed yet, that at the instant of the creation of the universe, what we call the big bang, there was only one primordial, fundamental force. As the universe expanded and cooled, this single force took on different forms until there were the four forms we have today. This is an example of one of the most important principles in physics today: *unification*. This idea says that many things that are apparently different are, in fact, closely related and are really just different forms of the same thing. We have just seen this with our discussion of forces, and we will see it again when we discuss the fundamental building blocks of matter: elementary particles.

How can we understand how one force can take on different forms? Well, we have seen a bit of this already when we discussed how the force between atoms can cause different forces, such as friction or wind force or muscular forces. We will also discuss, in the next chapter, the relation between the electric and magnetic force and how they are really manifestations of the single electromagnetic force. In addition, there is a very nice analogy that will also help us. Consider water. At very high temperatures, water has only one form: the gaseous form that we call steam. As the temperature cools, water can take on two very different forms, gaseous and liquid, which are very different from each other. Finally, as the temperature cools even more, water now can appear in three very different forms: gaseous, liquid, and solid. Yet they are all the same element—water. In an analogous way, as the universe cooled, the primordial force took on different forms.

Do we have any idea as to how this may have happened? The answer is yes. While we do not have the full story, we have been able to relate the electromagnetic force to the weak nuclear force and understand how these are different forms of a single electroweak force. We cannot go into the details here, but suffice it to say that we hope that by our understanding of the electroweak force we will gain some help in understanding how to merge the other forces. This quest to merge the fundamental forces goes back to Einstein, who thought there had to be a relation between electromagnetism and gravity. He spent the last 20 years of his life unsuccessfully trying to formulate what he called his unified field theory. We now know that gravity is the most difficult force to unify with the others. There are some theoretical physicists who believe that the way to do this is through a very esoteric theory known as string theory, where the fundamental objects in nature are minute (much, much smaller than a nucleus) vibrating strings in an eleven-dimensional universe. I said it was esoteric!

Figure 4.1 illustrates both the history of the unification of forces and the hoped-for-future. As it shows, the first unification was done by Newton when he showed that celestial (in the heavens) and terrestrial (on earth) gravity were the same. That

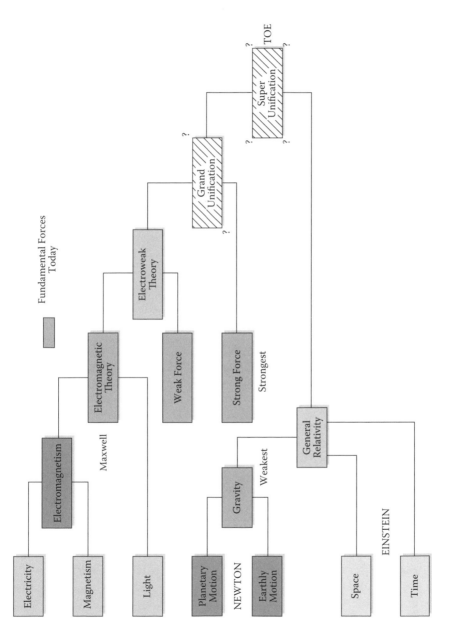

FIGURE 4.1 **(See color insert following page 102.)** The unification of forces.

is, he showed that it was the same force that kept the moon in its orbit that caused the apple to fall to earth. Just as Newton revolutionized the understanding of gravity in the seventeenth century, Einstein, some 230 years later, again caused a revolution in our understanding of gravity with his general relativity theory. But here, Einstein unified not two forces, but space and time into the single entity of "space-time." In this theory, gravity is caused by mass warping space-time. This, indeed, is a very different way of looking at a force.

Figure 4.1 also shows that electricity, magnetism, and light were unified by Maxwell, as we will discuss in the following two chapters. In addition, it shows the unification of the electromagnetic force with the weak nuclear force into the electroweak force. When we, hopefully, learn to unify the strong nuclear force with the electroweak, we already have a name for this merger: grand unified theory (GUTS). Finally, the grand unification of gravity with the others has been given the name TOE, representing the theory of everything. It is the hope of some physicists that string theory will be the TOE.

While we will not go any deeper into these esoteric ideas here, hopefully you are beginning to realize that our understanding of forces goes well beyond the simple description of a push or pull. It lets us get a glimpse as to the marvelous way nature has organized herself.

4.2 A SPECIFIC FORCE LAW: NEWTONIAN GRAVITY

While gravity is usually the first force that most people think of, have you ever wondered about what causes it or what determines how strong it is? This is usually given by a force law, which is an equation telling us how to calculate the size of a particular force. Many of you, I suspect, have already seen Newton's universal law of gravity. It is given by the equation that tells us the size of the gravitational attraction between two masses:

$$F = G(m_1)(m_2)/d^2 \tag{4.1}$$

G is known as the universal constant of gravity and has a value of 6.7×10^{-11} in units of N-m²/kg². This is a very small number and is one of the reasons that gravity is the weakest of all of the forces. The quantities m_1 and m_2 stand for the two masses that are attracting each other, and d is the distance between the two masses.

There are several things to note about this equation. First, it depends on the product of two different masses. A single mass does not produce a gravitational force. In fact, mass is the cause or source of gravity. So, mass is a very important quantity and we should make sure we understand what it is. Unfortunately, there seems to be a good deal of confusion between mass and weight. While they are related, they are not the same thing at all. Mass is a fundamental quantity like length or time, while weight is a force due to the earth's gravitational pull on an object on, or near, the surface of the earth. We will get back to this shortly. The best way to think about mass is that it is a measure of the amount of matter in an object, closely related to the number of atoms in an object. So, an object has the same mass anywhere in the universe. As we will see below, this is not true for weight.

The second quantity to discuss is the distance d. We should note that the force does depend on how far the objects are apart, and that the larger the distance, the smaller the force. In fact, since gravity depends on $1/d^2$, this tells us, for instance,

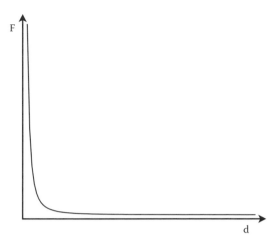

FIGURE 4.2 Force of gravity as a function of the distance between two masses. The farther the masses are apart, the weaker is their gravitational attraction.

that if we double the separation between the two objects, the force will decrease by a factor of 1/4. The graph in Figure 4.2 shows how the gravitational force caused by two masses varies as the distance between them is changed.

4.2.1 WEIGHT

As stated above, weight is the force on an object due to the gravitational attraction of the earth. We can use the universal law of gravity above to write an expression for the gravitational attraction between the earth, whose mass will be designated by the symbol M_e, and some mass, designated as m, on or near the surface of the earth. Before we can do this, we have to discuss what we use for the distance, d, since wherever the mass m is located, part of the mass of the earth is very close to m and other parts are very far away. It turns out when you average the effects of all the mass of the earth, it acts as if all the mass acts at the center of the earth. Thus, we use for the distance d, in this case, the radius of the earth, which we designate a R_e. We then write for the weight

$$W = \text{weight} = GM_e m/R_e^2 = m(GM_e/R_e^2) \tag{4.2}$$

We will get back to this equation very shortly to discover some very interesting consequences.

4.3 HOW DOES FORCE AFFECT MOTION? NEWTON'S SECOND LAW

It was stated at the beginning of Chapter 3, on motion, the best way of learning about forces is to study the effect they have on motion. The law that describes the relation between force and motion is Newton's second law of motion. It is usually written as

$$\mathbf{F} = \mathbf{ma} \tag{4.3}$$

This is a deceptively simple-looking relation, and also misleading as written. Usually when writing an equation, the implication is that the quantities on the right-hand side cause or determine the quantity on the left. For instance, in Equation 4.2, the weight of an object is determined by its mass and the other quantities in the parentheses. So, the way Equation 4.3 is written implies that acceleration causes force, but just the opposite is true. It is a force applied on an object that causes the object to accelerate. Thus, in order to convey the correct meaning, we should write Newton's second law as

$$\mathbf{a} = \mathbf{F}/m \qquad (4.4)$$

In addition, $\mathbf{F}$ can be easily misinterpreted. For instance, I assume you are probably reading this while sitting in a chair. If I were to ask you if there was a force acting on you at this moment, you would most likely say, "Yes, gravity is pulling me down." Equation 4.4 seems to imply that the gravitational pull should cause you to accelerate downward. But that is certainly not true, since you remain stationary and are clearly not accelerating. The solution to this apparent problem has to do with the meaning of the symbol $\mathbf{F}$. $\mathbf{F}$ is just not any force, but the sum of all forces acting on the object. The reason you do not accelerate while sitting in your chair is that there are two forces acting on you. One is indeed gravity pulling you down, but the other is the equal and opposite force of the chair acting upward. These two forces exactly cancel each other. So, the sum of these forces (sometimes we call the sum of all forces the net force) is zero. Equation 4.4 then predicts that the acceleration should indeed be zero.

There is one other subtlety in the meaning of $\mathbf{F}$. Only forces that are applied to an object externally can influence its motion. Forces that are internal (inside) an object cannot cause any change in motion. So, the full meaning of the symbol $\mathbf{F}$ is that it is the sum of all external forces acting on an object. If there were three different forces acting on an object, then we can write

$$\mathbf{F} = (\mathbf{F}_1)_{\text{ext}} + (\mathbf{F}_2)_{\text{ext}} + (\mathbf{F}_3)_{\text{ext}} \qquad (4.5)$$

where the subscript ext is there to remind us that the force must be external to the object. In other words, we only consider forces that act *on* the object.

Before continuing, let us go back to our discussion in Chapter 3 about the acceleration of a ball that is thrown into the air. If you remember, we said the acceleration must be a constant throughout the entire motion. In Chapter 3, it was argued that was correct because the graph of velocity versus time was a straight line. This meant the slope was constant, and since the slope of that graph is the acceleration, then the acceleration had to be constant. This was a graphical-mathematical argument, which, I suspect, was not that persuasive for some of you.

Now, using Newton's second law, we can also understand this by a physical argument. Equation 4.4 tells us if the net force on an object is constant, then the acceleration must be a constant. While the ball is in the air, the only force acting on it is the gravitational pull of the earth (we will neglect air resistance, which should be small and negligible). Near the earth (less than about 100 miles above the earth), the gravitational pull is constant and acting downward. Thus, the acceleration of the ball must be constant and in the downward direction.

4.4 NEWTON, THE APPLE, AND THE MOON

As we indicated above, Newton is arguably responsible for the first great unification in physics. Before him, it was thought that there were two different types of gravity: terrestrial gravity affecting objects close to the earth, and celestial gravity for objects far from earth. What Newton realized was that there was only one type of gravity. Thus, the force that caused the apple to fall from the tree was the same as that which caused the moon to move in its orbit. The motion of both objects was due to the earth's gravitational force on them. This has to be one of the great "ahas" in history. Think a moment about that—relating a falling apple to the moon. If that is not one of the most creative ideas in the history of thought, I do not know what is.

The fact that the gravitational force depended on $1/d^2$ is an important part of the story. What Newton realized was that it was the earth's gravitational pull on both the apple and the moon that caused them to accelerate. In the case of the moon, it is a centripetal acceleration (v^2/r, where r is the distance to the moon from the center of the earth) due to its circular motion. Newton could calculate this acceleration. The acceleration of an apple, or any other object near the earth, had been measured by Galileo and others. What Newton showed was that the ratio of these two accelerations was the same as the ratio squared of the earth's radius to the distance to the moon, confirming that the gravitational force had to fall off as $1/d^2$.

4.5 COMBINING TWO LAWS

One thing physicists do quite often is combine two or more theories to see if this leads to new understanding or predictions that could not be obtained from any one of them alone. In this chapter, we have been considering two of Newton's laws, his universal law of gravity and his second law of motion, which tells us how a force applied on an object will influence the motion of that object. Let us see what happens when we combine Equation 4.2 for weight (the gravitational force on a mass due to the earth) with Equation 4.4. In order to have the only force acting on the object be its weight, it will have to be freely falling in air. If it were resting, for instance, on a table, then there would be a second force on it due to the table. We will also neglect any effect due to things like air friction. So, in this case, there is only the single force of gravity acting on the object, whose mass we will designate as m. Using Equations 4.2 and 4.4, we can write

$$a = F/m = (GM_e m/R_e^2)/m = GM_e/R_e^2 \qquad (4.6)$$

There are several important things we should notice about this result. Before reading on, see if you can see them.

One is that the acceleration due to the earth's gravitational pull on the object does not depend on its mass—the mass has canceled out since it appears in both the numerator and denominator. In fact, since we did not specify the object or its mass, this result, namely, that the acceleration due to gravity does not depend on the mass, is true for any object of any mass. Also, since the acceleration, a, depends on a combination of three constants, it must be constant. This is an important result, so let us restate it clearly:

All objects falling only under the influence of gravity will fall with the same acceleration, independent of their mass.

At this point you may be saying, "Wait just a minute! I know very well from all sorts of experience that, for instance, a penny will fall a lot faster than a sheet of paper." Well, we have to pay careful attention to the words "*only* under the influence of gravity." In the case of the sheet of paper, because of its large surface area, air resistance cannot be ignored. There are two forces acting on the paper, one of which is acting in an upward direction, partially canceling out the downward force of gravity. In the case of the penny, there is also air resistance, but it is much smaller due to the much smaller surface area, so its effect can indeed be neglected. In fact, to prove this, I suggest you do the following simple experiment. Take a piece of paper from your notebook and drop it at the same time you drop a penny, or other coin. You will, of course, note that the coin reaches the floor well before the paper. Now, take the same piece of paper and crumple it into a ball. Drop the crumpled paper with the coin as before. Hopefully you will see a very different result.

This acceleration of objects due to gravity has been given the special symbol g. Many of you are probably familiar with it already. We see from above:

$$g = GM_e/R_e^2 = 9.8 \text{ m/s}^2 = 32 \text{ ft/s}^2 \qquad (4.7)$$

The weight of an object, using Equation 4.2, can then be written as

$$W = mg \qquad (4.8)$$

The weight of something can be easily measured by placing it on a scale. In fact, because weight can be so easily measured, it is the most common way of determining the mass of an object. The other method of measuring the mass requires doing an experiment. If we rewrite Newton's second law as $m = F/a$, we can determine m by applying a known force on the object and measuring the acceleration caused by that force. Just placing an object on a scale and reading the weight is obviously a good deal easier.

4.6   THE MASS OF THE EARTH

We have measured the mass of the earth to be 6×10^{24} kg.

Do you have a question here? This statement should cause you to stop and think and evoke a question in your mind. Before reading on, see if you can think of the question.

How do we measure the mass of the earth? After all, we cannot put it on a scale or apply a force to it and then measure its acceleration. So, how do we do it? Well, if we look at Equation 4.7, where we have defined g, we see the mass of the earth is part of that equation. We just need to know the value for all of the other terms in that equation. We do know g—Galileo measured it, and it is also measured by students in almost every first-year physics laboratory. We also know the radius of the earth, which is about 4,000 miles. In fact, this was first measured by the ancient Egyptians. But what about G? While Newton introduced it in 1685, he had no idea of its value. Without knowing the value of G, we have no way of measuring the mass of the earth or, for that matter, the mass of any other heavenly body, such as the sun.

It wasn't until 1798 that Henry Cavendish performed a very clever experiment to measure G. The idea was to use Equation 4.1, Newton's law of gravity. If he could measure the gravitational force between two known masses at a known distance of separation, then inspecting Equation 4.1 shows that the only unknown is G. In principle, this sounds very easy. In practice, it is very difficult because the gravitational force between any laboratory-sized objects is extremely small. Cavendish had to invent a very clever method to measure the exceedingly small force. Figure 4.3 depicts the experimental apparatus. The two large masses, M, are used to attract the two smaller masses, m. The smaller masses are connected by a rigid bar, which is suspended by a fiber that can twist. Initially the two larger masses are absent and the bar and smaller masses are just hanging by the fiber. When the larger masses are brought into position, the gravitational force attracts the smaller masses to the larger ones, causing the fiber to twist and rotate through a small angle as depicted in the figure. The size of the angle is directly proportional to the gravitational force. Because the force is so small, the angle is very small, and hence hard to measure accurately. Here is where the genius of Cavendish comes in. He found a way of measuring the

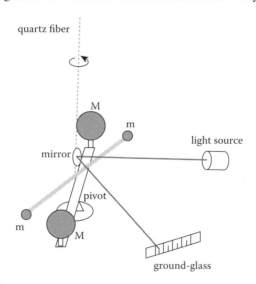

FIGURE 4.3   The Cavendish experiment. (From M. Simon; www.physics.ucla.edu/demoweb/demomanual/mechanics/gravitational_acceleration/cavendish_balance.html)

angle using what is known as an optical lever. Attached to the fiber was a mirror with a light source shining on it. Before the large masses were brought into position, the position of the light reflecting off the mirror was noted on a screen that could be placed a fairly large distance from the mirror. When the large masses were placed in position, the smaller hanging masses were attracted to the large masses, causing the fiber and mirror to twist by a certain angle. The spot of light gets deflected through twice that angle. What is actually measured is the distance the spot of light moves on the screen. This can be quite large, and thus measured with high accuracy. The amount of twist is directly proportional to the gravitational force between the set of known masses. Once Cavendish knew the force between the masses, he could solve Equation 4.1 for G.

As a testament to his skills as an experimenter, the value Cavendish obtained in 1798 was within 1% of the modern value. By the way, to make sure the scientists at that time got the point and also to "jazz" up the title of his paper reporting on his measurement of G, Cavendish entitled the paper "Weighing the Earth."

The Cavendish experiment very nicely demonstrates two important aspects of experimental physics. One is the creativity that is often necessary to find ways to do the desired measurement. The other is the indirect nature of experimental evidence. In this case, in order to find G and the mass of the earth, the measured quantity was the distance the spot of light moved.

Once we knew G, we could also use the moon to get the mass of the earth. In a similar manner, we could also get the mass of the sun. Can you think about how this could be done?

4.7 NEWTON'S FIRST LAW

We have already discussed Newton's second law. His first law is closely related to the second law. It is known as the law of inertia since if something has a lot of inertia, then it is resistant to change. The first law is usually stated as:

A body in straight-line motion tends to stay in straight-line motion, and a body at rest tends to stay at rest unless acted on by an outside force.

It states that things will not change unless forced to do so. This holds true for inanimate objects as well as people. There is another, much more efficient way of saying the same thing:

A body will maintain constant velocity unless acted on by an outside force.

Or:

The acceleration of a body will be zero unless acted on by an outside force.

This last statement sounds like the first law is just a special case of the second law, which, remember, states: a = F/m. If F is zero, then a must be zero. On the other hand, the first law can be used to define the action of a force: a force is something that causes a change in velocity. The second law then tells how big the effect of the force is. Let us look at a familiar situation to get a better understanding of the first and second laws.

4.7.1 What and Where Is the Force?

When riding in a car and going into a circular turn, think about what you experience. What you will probably say is that you feel a force pulling your shoulders toward the outside of the turn. Let us analyze this carefully. If there is indeed a force pulling on you outward, then what is its cause? In other words, what is the physical agent producing the force? After all, since we have defined force as the interaction between two objects, then what is the other object producing the outward force on you? For instance, if a rope was attached to you, the other object would be the rope. I think you will find that it is very difficult to identify the other object for the very good reason that there is no other object.

So, what is going on? Part of the answer is that you are getting somewhat fooled by the first law. First, there *is* a force on you due to the fact that you are attached to the car, either by a seatbelt or the friction between you and the car seat. But this force is pulling you with the car, causing you to accelerate in a direction toward the center of the circle (remember our earlier discussion about circular motion). This is an inward force, not in an outward direction. The reason you think you feel an outward force on your shoulders is partly due to the fact that we are not rigid objects, but are quite flexible. In addition, the first law states that objects, because of their inertia, want to maintain their straight-line motion. Thus, while the bottom of your body is being pulled inward, your shoulders want to continue in the original straight-line direction. But as we discussed in the previous chapter, the straight-line direction is outward from the center of the circle. So, it appears that your shoulders are being pulled outwards due to their tendency to go straight.

To summarize, Newton's second law says since you are moving in a curve and hence accelerating inward, there must be an inward directed force acting on you. The first law says that your shoulders, being loosely connected to the bottom part of you, would rather go straight, which feels outward relative to the center of the curve.

4.8 NEWTON'S THIRD LAW

The third law, in distinction to the first two, does not say anything about motion. It does, however, relate certain forces to each other. In fact, it explains why only external forces can cause an object to accelerate. It states:

For every applied force there is an equal and opposite reaction force.

The applied force is sometimes called the action force, while its partner, the opposite force, is called the reaction force. Note that these action-reaction pairs act on different objects. If the chair you are sitting on applies an upward force on you, then you must apply a downward force on the chair. Understanding this, do you see why only external forces can cause an object to accelerate?

4.8.1   How Does a Horse Pull a Wagon?

We see the third law specifically relates these action-reaction pair of forces in that they must be equal and opposite to each other. While this sounds like a fairly straight-forward statement, it seems to engender a good deal of confusion. Consider, for example, a horse pulling on a wagon. The third law states that the wagon must pull on the horse with an equal and opposite force that the horse applies to the wagon.

> Do you have a question here? How, then, can the horse or wagon ever move since these two forces must cancel out. Can you see the solution? Think about this before reading further.

The answer is that the action-reaction forces act on different objects, not on the same object. Neglecting friction, in the axle of the wheel, the only force acting on the wagon is the pull due to the horse. This is what causes the wagon to accelerate. On the other hand, there are two forces acting on the horse. One is the frictional force between the horse's hoofs and the ground—in this case, we certainly cannot neglect friction. In fact, it is the force that produces the motion. The second force is that of the wagon acting on the horse pulling the horse backward. This is the reaction force of the horse pulling on the wagon. The frictional force between the hoofs and the ground must be larger than the one due to the wagon pulling on the horse, or there would not be an imbalance of forces allowing the horse to accelerate forward.

4.8.2   How Can We Walk?

You may now be wondering how a frictional force can cause the horse to accelerate. After all, doesn't friction slow things down? In order to answer this, think about how you walk. Well, without worrying about all the detailed biology, the brain sends a signal to your leg muscles to contract and push. It sounds pretty straightforward, except for one rather large problem. When we first discussed Newton's second law, it was pointed out that only forces external to a body could cause that body to acceler-ate. Certainly your muscles are inside of you. So, how can they cause you to move?

> Again, before reading further, try to answer this. One hint: Newton's third law.

Your muscles do indeed contract and cause your foot to push on the floor, but backwards (check this out by noticing the direction your foot pushes when you want to start to walk). The reason your foot exerts a force on the floor is due to the frictional force between your foot and the floor. But this is still a backward push. Here is where the third law comes to the rescue. As your foot pushes on the floor backward, the floor, having carefully studied Newton's third law, provides the external force that pushes your foot and the rest of you forward, causing you to accelerate forward. So, when you are ready to get up from your chair and walk somewhere, thank the floor for pushing you around.

Finally, we can understand why internal forces cannot have any effect on the acceleration of an object. Let us consider two interacting atoms in an object. Atom 1 produces a certain force on atom 2. But atom 2 produces an equal and opposite force on atom 1. Since both atoms are inside the object, the net force they produce on the object is zero. If this is true for any two atoms in the object, it is true for all pairs of atoms in the object. Hence, the total effect of all internal forces in an object always adds up to zero.

4.9   END-OF-CHAPTER GUIDE TO KEY IDEAS

- What are the fundamental forces in nature?
- How are most common forces related?
- What is the history of the unification of forces up to this date?
- What hoped for unifications have we not been able to do yet?
- What is the equation for the force of gravity, and what does it tell you?
- What is weight?
- What is Newton's second law of motion, and what is its meaning?
- What is the definition of **F** in Newton's second Law?
- What do we learn when we combine Newton's second law with the law of gravity for an object falling near the earth?
- How can we determine the mass of the earth? What else has to be known, and how is it determined?
- What is Newton's first law? What is Newton's third law?
- Can you explain how we walk?

QUESTIONS/PROBLEMS

1. Define *force*.
2. What is the difference between centripetal and centrifugal force? Give examples of both.
3. What is meant by "the unification of forces"?
4. List the four fundamental forces. Which of these have we now unified?
5. Why are the following forces due to the electromagnetic force: friction, contact, muscular, and chemical?
6. What are the dominant fundamental forces at work:
   a. When there is friction
   b. When a bat hits a ball
   c. When a balloon floats in air
   d. When a piece of wood floats in water
   e. When a knife cuts butter
7. Write Newton's law of gravity and explain, in words, the meaning of each term.
8. The distance between two masses is doubled. What happens to the force between them?
9. What is the difference between mass and weight?

10. Describe two different ways of measuring mass.
11. Write Newton's second law in the form that conveys its correct meaning. Explain why this is the way it should be written.
12. Define the symbol **F** as it appears in Newton's second law.
13. Is the earth accelerating? Justify your answer.
14. We have proved in this chapter that all objects near the earth fall with the same acceleration. If this is so, explain why a feather falls so much slower than a penny.
15. Explain why Cavendish entitled his paper "Weighing the Earth."
16. What did Cavendish actually measure in his experiment? What did this allow him to calculate?
17. What was the clever experimental technique Cavendish employed that allowed his experiment to be successful?
18. Why is Newton's first law also known as the law of inertia?
19. In going around in a circle, in your car, for instance, list all the forces acting on you and their direction. Also, list the corresponding reaction force.
20. Explain why only forces external to an object can cause that object to accelerate.
21. Choose which statements below are true or false. For those that you have chosen as false, explain why and correct them to make them true. When a ball is thrown into the air (neglect any air friction and only consider the motion while it is in the air, not in contact with the hand):
    a. The acceleration is greatest right after it has been released from the hand on the upward throw.
    b. The acceleration is greatest just before it is caught on the way down.
    c. The acceleration is positive as the ball moves upward and negative as the ball moves downward.
    d. The acceleration is negative as the ball moves upward and positive as the ball moves downward.
    e. The acceleration is always positive.
    f. The acceleration is always negative.
    g. The acceleration is zero at the top of the motion.
    h. The acceleration is greatest at the top of the motion.
    i. The acceleration changes sign as the ball moves through its highest point.
22. A compact car and a bus go through the same turn at the same speed. Fill in the blank line with either *less than*, *equal to*, or *greater than*.
    a. The acceleration of the car is _____ the acceleration of the bus.
    b. The force on the car is _____ the force on the bus.
    c. The force the car exerts on the road is _____ the force the bus exerts on the road.
    d. The force the car exerts on the road is _____ the force the road exerts on the car.
    e. The force on a twin in the car is _____ the force of the other twin in the bus.

23. Show how we can use what we know about the moon to measure the mass of the earth.

24. Show how we can use what we know about the earth to measure the mass of the sun.

25. Describe a situation of an object in motion such that at one time in the motion, gravity acts as a centripetal force, and at another time, it acts as a centrifugal force. (Other forces could be present.)

5 Electromagnetism

In the previous chapter, we discussed forces in general and the gravitational force law in particular. In this chapter, we will discuss the electromagnetic force in some detail. It is an extremely important force to us since, as was mentioned in Chapter 4, besides gravity, it is the only other force that has any direct effect on us.

5.1 THE ELECTRIC FORCE LAW

The electric force law is similar, in many ways, to Newton's law of gravity. It is given by the equation

$$F = Kq_1q_2/d^2 \tag{5.1}$$

where q_1 and q_2 are electric charges that are measured in units of coulombs. There are two different types of charges, positive and negative. The unit of charge is named after Charles Coulomb, who first wrote down Equation 5.1, which is, in fact, known as Coulomb's law. Charge is a fundamental property of matter. It is independent of the other fundamental properties of mass, length, and time. Thus, the coulomb is a fundamental unit, not expressible in terms of the other units. The charge of the electron or proton, which is the smallest observed charge, is 1.6×10^{-19} coulombs. Some of you may have heard about quarks, which are believed to be the basic building blocks of matter (protons are made up of three of them) and have one-third or two-thirds the charge of the electron. But free quarks are never observed directly, so the statement above about the smallest observed charge is still true. We will discuss quarks and other basic building blocks in a later chapter.

K is a universal constant, like G, but its value is 9×10^9, compared to the much smaller value of 6.7×10^{-11} for G. The units of K are N-m²/(coul)². The quantity, d, has the same meaning as in the law of gravitation. It is the distance between the two electric charges.

The fact that the gravitational force law and the electric force law are so similar caused Einstein to believe that the two laws were intimately connected and were, in fact, different manifestations of the same force. As mentioned earlier, he spent the last 20 years of his life looking for what he called the unified field theory. To get some idea as to why it is so difficult to unify electromagnetism and gravity, let us consider the following major differences. First is the fact that gravity is only attractive while the electric force can be either attractive or repulsive. We all learned a long time ago that unlike charges (+, − or −, +) attract while like charges (−, − or +, +) repel. How can two forces be related if one has repulsion while the other does not?

The second major difference has to do with the relative strengths of the two. The electromagnetic force, relative to gravity, is very much stronger. To see this, let us consider the ratio of the two forces between two fundamental particles, electrons. If we use Equation 5.1 from this chapter and Equation 4.1 from the previous chapter, we can write down that ratio

$$F_E/F_G = Ke^2/Gm_e^2 \approx 10^{42}$$

where $e = 1.6 \times 10^{-19}$ coulomb, the basic charge on the electron, and $m_e = 9.1 \times 10^{-31}$ kg, the mass of the electron. Given the values of K and G, you can work this out yourself. (Note that in the ratio, the distance d does not appear since it cancels out.) Let us make sure we grasp the meaning of this ratio. It is an extremely large number, one of the largest known in science. It tells us how much larger, in some intrinsic way, the electric force is compared to the gravitational force. We can therefore ask a similar question as above: How can these two forces be related if one is so much larger than the other?

These two questions give some feeling as to why it has been so very difficult, as Einstein found out, to unify gravity and electromagnetism. When and if we do find the unified theory, it will give us the answers to these questions.

5.2   UNIFYING ELECTRICITY AND MAGNETISM

From a historical point of view, it could be argued that the first unification of forces was done by Newton with gravity. Before then, it was thought that there were two different types of gravity: celestial and terrestrial—one for the heavens and one for earth. Part of his creative genius was to realize that there was no difference. The force that caused the apple to fall was the same one that kept the moon in its orbit around the earth. But one could argue that this dichotomy before Newton was due more to philosophy and ignorance than to some known physical differences. On the other hand, electricity and magnetism seem to be physically quite different. For instance, magnets are bipolar, which means every magnet must have both a north pole and a south pole. If you break a magnet in half, you create two magnets, both with north and south poles. On the other hand, while electric charges also come in two varieties, they are separate, or unipolar. We can have only positive charges or only negative charges, or any combination.

5.2.1   Ampere's Law

You may have heard of Ampere before, or at least his name, since it is the unit of electrical current. What Ampere did was to discover one of the basic relationships between electricity and magnetism. His law, simply stated, is:

Moving charges (electric currents) produce a magnetic field.

You should be very familiar with many applications of this law since it is the basis for all electromagnets that are used in telephones and loudspeakers, to name just a few. We will get back to a further discussion of Ampere's law shortly.

You may have noticed that a new concept, magnetic field, was introduced above. In the next section, the phrase *electric field* is introduced. You may already be familiar with the idea of a field and be comfortable with it. If not, you may want to jump ahead to the next chapter, where the field idea will be discussed in detail.

5.2.2 FARADAY'S LAW

Ampere's law states that an electrical quantity can produce magnetism. Faraday's law tells us how magnetism can produce electrical quantities. It states:

A changing magnetic field creates an electric field.

It is the principle upon which electric generators work, and hence is responsible for the vast majority of our electric power generation. In any electric generation plant, a large turbine wrapped with conducting wires rotates at high speed in a magnetic field. The rotating wires experience a changing magnetic field, which causes an electric current to flow in the wires according to Faraday's law. This current is then transmitted for our use by high-tension wires to our homes, businesses, schools, etc. Energy is needed to cause the massive turbines to rotate. This energy can be supplied by a variety of sources: coal, oil, gas, nuclear, or falling water. Where there is a steady wind, a farm of windmills can be used to turn smaller turbines. While the sources of energy may be quite different, in all of these cases, it is through Faraday's law that the electricity is produced. The only other way to generate electricity is by solar power, which directly converts sunlight to electricity. Only a small fraction of our power needs has been generated in this way.

5.2.3 THE LORENTZ FORCE

To be complete, there is another connection between electricity and magnetism. Earlier in this chapter we talked about Coulomb's law, which describes how one charge produces a force on another electric charge. But there is another way to produce a force on an electric charge. That occurs when a charge is moving in the presence of a magnetic field. This is known as the Lorentz force. It depends on the speed of the charge and the direction in which it is moving with respect to the magnetic field. It has many applications. For example, in your television tube, it is used to both bend and focus the electron beam that, when it hits the screen, forms the picture you see. We will not say too much more about the Lorentz force, but it is another example of the intimate connection between electricity and magnetism.

5.2.4 BACK TO AMPERE'S LAW

Ampere's law was stated above as a simple, single-line statement. Hopefully it is clear why it is the basis for the electromagnet, a very useful device. But now let us state it in a stronger form:

**The *only* way to produce magnetic fields (magnetism) is by
moving electric charges.**

Do you have a question here? The statement of Ampere's law above should
evoke a question in your mind. Hopefully you formed the question before reading this far. If not, look at the statement again and see if you can come up with
a question before reading further.

Here is the question: If only moving charges can cause magnetism, what causes
a bar magnet, or any other type of permanent magnet (usually made of iron), to be
magnetic? In other words, where are the moving charges in the permanent magnet?

So, we have the question; now see if you can come up with the answer
before reading further.

By the way, this is basically how science works. We notice something that we do not
necessarily understand. This evokes a question, which we then try to answer. When
we find the answer, we find something new about nature. Our answers usually come
in the form of guesses, which we then check to see if they make sense or hold up to
experimental evidence.

OK, let us do some guessing. Some of you may have thought that perhaps the
north and south poles have opposite charges that somehow jump from one pole to
the other, thus providing the moving charges. To test this idea, we could try putting
some material in the way, a piece of plastic, for instance, which would prevent the
charges from getting through. If we did this, we would find out that the magnet still
had the same strength as before. So, while this is certainly a reasonable guess, the
experimental evidence tells us it is not correct.

Others of you may have thought that there is an electric current flowing in the
iron from one pole to the other. Note this is different from the first guess, which had
charges jumping off one pole and going through the air to the other. The main problem with this is that if there was indeed a current in the iron, let us say positive charges
flowing from the north to south pole, then there would be a pile-up of positive charges
on the south pole. Eventually (very quickly, actually) the positive charges would repel
other positive charges and the current would have to stop and the iron would no longer
be magnetic. But we know permanent magnets are just that—permanent.

5.2.5 WHERE ARE THE MOVING CHARGES?

Let us think carefully about what the magnet is made of. Most magnets are made of
iron. What do we mean when we say something is iron? What we finally mean is that
it is made up of iron atoms. And what is an atom made of? We discussed this briefly
earlier when we mentioned the planetary model of the atom. An atom is composed

of electrons moving in a circular orbit around the nucleus. So, here are our moving charges: the electrons in the atom.

> Do you have another question here?
> Hopefully the statement above has evoked another question in your mind. If not, again, before reading further, try to think about why you should be disturbed by that statement.

Here's the question: Every atom has moving electrons, so why aren't all atoms, and hence all materials, magnetic? The answer is not simple and takes a detailed understanding of atomic structure to answer fully. A hint comes from the periodic table and where the magnetic materials, nickel, cobalt, and iron, are located in it. They are right next to each other, which means they are somewhat similar in their atomic structure. We do not have to go into the details here, but if the question was not asked, then scientists would not have looked for the answer, which has led us to a much more detailed understanding of the atom and magnetic materials.

For those of you who are interested, and to be complete, there is a detail about our magnetic story that should be explained. It was implied above that it is the motion of the electrons in their orbit around the nucleus that produces the magnetism. In reality, there is another aspect of the electrons' motion that is responsible for the magnetism. Every electron also spins around an axis, just like the earth or a spinning top. Since electrons are charged, they act as spinning (thus moving) charges, and hence produce a magnetic field. In fact, every electron acts like it is a tiny magnet. In the three magnetic materials mentioned above, these tiny magnets can all align along a certain direction. While each electron is a minute magnet, there are so many of them that their combined effect can produce the macroscopic magnets that are used in so many applications.

5.3 END-OF-CHAPTER GUIDE TO KEY IDEAS

- What is Coulomb's law, and what does it tell us?
- What is Ampere's law?
- What is Faraday's law?
- What is the Lorentz force?
- What do the above three have to do with the unification of electricity and magnetism?
- What causes magnetism in a bar magnet?

QUESTIONS/PROBLEMS

1. Who "unified" gravity and how did he do it?
2. Discuss the ways that the electric and gravitational forces are similar and how they are different.
3. Give two reasons why it is difficult to unify electricity and gravity.

4. What produces the electric force?
5. How do electricity and magnetism differ?
6. List and state three laws that relate electricity and magnetism. State which specific devices are based on which specific law.
7. What is the cause of magnetism in a bar magnet?
8. What is the basis for stating that the electric force is so much greater than the gravitational force?
9. Since the electric force is so much greater than the gravitational force, why do we say that it is gravity that holds us to the earth?
10. A typical book may contain 10^{25} charged particles.
    a. Why don't we notice any electrical force between two nearby books?
    b. Why don't we notice any gravitational force between the books?

6 The Field Concept

6.1 WHAT IS THE CONNECTION?

Consider the following situations: a charged rod attracting or repelling a charged balloon, a magnet attracting or repelling a compass needle, a ball either thrown or released falling to the ground, the moon kept in an orbit around the earth, and the earth in orbit around the sun. What do all of these things have in common? Again, think about this.

For one, all these situations involve forces, either attractive or repulsive. But in addition, there is something special about these forces. Do you see it? In all of these cases there is no *contact* between the interacting objects. Have you ever wondered how objects interact with each other over some distance? For instance, how does the earth attract a ball that is in the air, or how does the earth attract the moon about 250,000 miles away, or better still, how does the sun attract the earth at a distance of 93 million miles? In other words, how do noncontact forces work? That is what we are going to try to answer in this chapter. After all, we are all very familiar with contact forces. You are touching this book right now; you are touching the seat you are sitting in, and your feet are touching the floor. Nothing seems particularly mysterious about these "touching" forces. It is probably the noncontact forces that seem strange. Well, by the end of this chapter, there will be some surprises in store for you.

6.2 ACTION AT A DISTANCE

In considering noncontact forces, we will use the electric force law as our example. But what we conclude will be equally true for any other force. So, let us remind ourselves of Coulomb's law, the electric force law, as expressed in Equation 5.1 of Chapter 5:

$$F = Kq_1q_2/d^2 \qquad (6.1)$$

Remember, d is the distance between the two charges, and d could be any distance.

The first explanation we will consider is known as action at a distance. It states:

It is the *nature* of charged objects to have a force on them as given by Equation 6.1.

At this point you may be thinking that this really is not very explanatory. It sounds much like an Aristotelian explanation, similar to "it is the nature of objects to fall to the earth," or "it is the nature of earth-bound objects to slow down and stop."

6.2.1 Is This a Legitimate Explanation?

This is really the same question as: Is this a legitimate theory? We have discussed this in Chapter 2. Remember, a theory must be predictive. So, the question is: Is action at a distance predictive? The answer is yes.

Equation 6.1 tells us that the force is dependent on the distance. If the distance between the two charges is instantly changed, then the force on either charge will be instantly changed, with no delay in time. So, we do have a definite prediction that can be tested. If one charge is moved very rapidly, let us say away from the other, then the other charge will instantly feel a smaller force. No matter how far the charges are initially apart, the change of force will be felt instantaneously. Before discussing this further, let us consider the other possible theory.

6.3 THE FIELD CONCEPT

In order to appreciate one aspect of the field concept, let us consider aother aspect of action at a distance. Note that in the latter idea, two charges are required to produce a force. In other words, there is no difference whether there is one charge or no charge present. The field idea actually makes a distinction. It states that a single charge that is present in a region of space has an effect on space. Be careful, it does not say a single charge produces a force. This would indeed be incorrect since Coulomb's law (and experimental evidence) says that two charges are necessary to produce the electric force. This effect on space is, in fact, the electric field. The electric field is produced (created) by a single charge and emanates throughout space. In other words, a single charge affects (warps) space. It is a rather abstract idea.

The following analogy may be helpful. Consider a flat mattress. If we place something heavy on it, like a bowling ball, the shape of the mattress will be warped and will no longer be flat. In an analogous way, a charge creates an electric warping of space.

Hopefully, the diagram below will help you visualize it. The circle in the center represents an electric charge, call it q_1. The arrows represent the electric field produced by q_1, call it E_1. The outer circle represents another charge, q_2, in the vicinity of q_1.

6.3.1 How Does This Help Explain Noncontact Forces?

These arrows just represent some of the electric field lines. The electric field actually permeates all of space. The reason the field is represented by arrows is that the field has a direction as well as a strength, so it is a vector. In Figure 6.1, the field points away from charge q_1, indicating that q_1 is a positive charge. If q_1 were negative, the field would point inward instead of outward. For the moment, let us just concentrate on q_1. The idea is that q_1 alone has an effect on space that is the electric field. The field permeates throughout all of space, thus extending well beyond q_1 itself. The strength of the field falls off as the distance from q_1 increases. In fact, the field strength is given by the equation

$$E_1 = Kq_1/d^2 \tag{6.2}$$

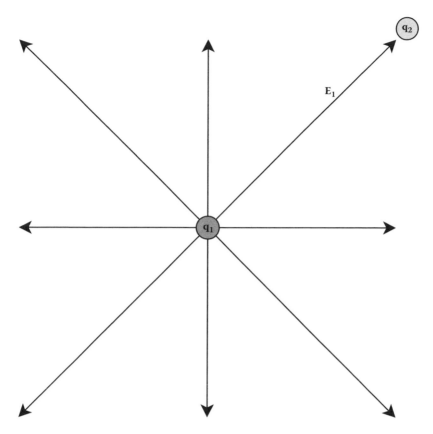

FIGURE 6.1 Electric field lines produced by charge q_1. The electric field can then act on another charge q_2 at a distance d away from q_1.

We can note several things about this equation. First, it depends only on the charge q_1 since it is q_1 that produces it. Second, it diminishes with distance as $1/d^2$. It is important to realize that in this equation, d is *any* distance from q_1. It represents the distance to any arbitrary point in space. Nothing has to be at this particular point. This is because we are talking only about how q_1 affects space. Now let us bring in q_2. If we place a second charge, q_2, at some distance, d, from q_1 as indicated in Figure 6.1, q_2 will be in the presence of the electric field E_1.

Here is where the force is produced.

In the field idea, the force on q_2 is caused by the action of the field E_1 acting on q_2. It *is not* q_1 acting directly on q_2 that produces the force on q_2, rather it is the *field* produced by q_1 acting on q_2 that causes the force on q_2. Imagine, if you will, that we could "freeze in" the field E_1 due to q_1 such that even if we remove q_1, E_1 would still be there (this cannot happen, of course, but let us just imagine that it could happen). If this were possible, then the other charge, q_2, would still have a force on it due to it being in the presence of the electric field, E_1. This is very different from action at a distance, where the force is produced by the direct action of q_1 on q_2.

Using our mattress analogy above, if a marble were to be placed on the warped mattress, it would move toward the bowling ball, not because of the direct attraction of the marble to the ball, but due to the fact that the space it was in (the mattress) was indented (warped) and the marble would move according to the shape of the mattress.

To review, the chain of reasoning is that a charge produces an electric field. The field produced by that charge is present throughout space. If there is a second charge somewhere in that region of space, then the field produces a force on that second charge. One thing should be clarified. *Any* charge produces its own field. Above, we have talked about the charge q_1 producing a field. Of course, the charge q_2 also produces its own field. Also, fields *do not* act on each other. Fields only act on charges.

For those of you who are interested, and to be complete, Equation 6.3 tells us how the field and force are related:

$$F_{(on\ q2)} = q_2 E_1 \qquad\qquad (6.3)$$

What this says is if there is a charge q2 in the presence of a field E_1 produced by charge q_1, then there will be a force on q_2 as given by Equation 6.3. You should also note that if we combine Equations 6.2 and 6.3, we get back Equation 6.1, the Coulomb electric force law of Chapter 5.

$$F = q_2 E_1 = q_2 (Kq_1/d^2) = Kq_1 q_2/d^2$$

6.3.1.1   Thinking Like a Physicist

At this point you may have several thoughts going through your mind: this is complicated; it is abstract; what kind of mind comes up with ideas like this? But if you are at all beginning to think like a physicist, you would have the following question: If the field exists, how do we detect it?

In our day-to-day lives, the way we normally detect things is through direct sensory perception. But as we shall see many times in this book, in science, it is usually through indirect means that we finally "observe." In the case of the field, we cannot see, feel, taste, hear, or smell it. In fact, the only way to detect the field is to see if there is a force produced on another charge particle. If there is, then we say there must be an electric field present.

Now, you should really be up in arms. After all, we have introduced the field idea to explain forces, but now you are being told that the only way to detect the field is to see if there is a force. It sounds like a circular argument and, to make it worse, pretty complicated. How is this any different than action at a distance?

6.3.1.2   Is There a Way to Tell the Difference?

If we restrict ourselves to consider only stationary charges, then, in fact, the field idea and action-at-a-distance predictions are exactly the same. But if we consider moving the charge q_1 rapidly away from q_2, as we did above when we discussed action at a distance, then the field concept predicts something very different. Namely, it predicts that there will be a time delay until the charge q_2 feels the new, diminished force. This time delay will depend on the distance between the two charges. The farther the charges were initially apart (the larger the distance d), the larger the time delay will be. This, then, is very different than the prediction of action at a distance. It is the type of situation physicists love since we can do an experiment that will unambiguously test which is correct. When we perform the experiment, we see that the field prediction is indeed correct.

While it is somewhat difficult to display an electric field, it is fairly easy to see a magnetic field. In fact, many of you may have seen such a demonstration before when iron filings have been sprinkled over a magnet. The iron filings form a pattern displaying the shape of the magnetic field lines.

6.3.2 UNDERSTANDING THE TIME DELAY

To understand why the field concept requires a time delay, consider Figure 6.2. It shows the same electric charge in two different positions. On the left, the charge has been stationary and one of its electric field lines is shown pointing away from it. The charge is rapidly moved (accelerated) so at some instant it is at the second, right-hand position. Near the second position, the corresponding field line must point away from where the charge is now. But the far-away field does not yet know that the charge has been accelerated and is still pointing away from the original position. Since field lines must be continuous, the dashed kink shows how the old and new field lines must be joined. The kink is a readjustment (change) of the field due to the acceleration of the charge. It propagates outward with a definite, finite speed, thus

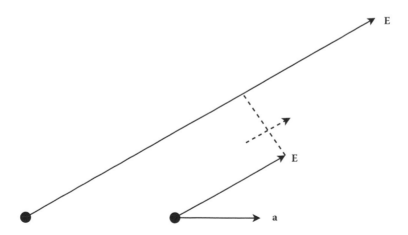

FIGURE 6.2 Left-hand stationary charge with one of its electric field lines as indicated. The charge has now been accelerated to its right-hand position with its corresponding field line also indicated. The dashed kink joins the later field line (from the accelerated position) with the earlier one, telling it that the charge has been accelerated. The dashed arrow indicates that the kink is moving in the direction indicated. The arrow labeled a indicates that the charge has been accelerated.

taking some time to inform the distant field there has been a change; therfore, there must be a time delay that depends on distance.

6.3.2.1 The Speed and Identity of the Kink

As stated above, the kink travels at a definite speed. That speed is 186,000 miles/s, or 3×10^8 m/s. Some of you may recognize this as the speed of light. In fact, that is exactly what light is: the readjustment of the electric field when charged particles are accelerated. So, not only does the field concept predict the correct time delay, but it predicts and explains what light is. To emphasize this point, light is only produced when charged particles are accelerated. We will discuss the nature of light in a great deal more detail in Chapter 11, where we will see that light is an electromagnetic wave that depends on both changing electric and magnetic fields.

6.4 BACK TO CONTACT FORCES

We began this chapter trying to understand noncontact forces. These seemed a bit mysterious, especially compared to the more familiar contact forces. It is now time to take a close look at these "touching" forces we think we understand.

First, let us make sure we have a clear definition as to what we mean when we say objects are in contact. A clear and unambiguous definition is that objects are in contact when the distance between them is zero (d = 0). Now let us carefully consider what is touching when we say, for instance, my hand is touching the table.

Think about this before reading further. What is really touching what?

What we really have to do is look at the hand and table on the *microscopic* scale. If we do that, we have to consider the *atoms* that make up both the hand and table.

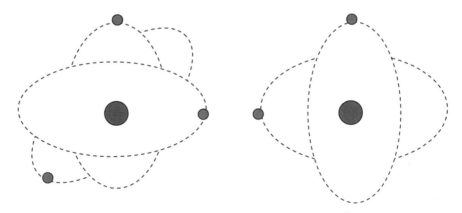

FIGURE 6.3 (See color insert following page 102.) Two atoms (hand and table) close to each other. The dashed ellipses represent the electron (blue) orbits that circle the nucleus (red circles). This is not to scale since the size of the electron orbits is about 100,000 times the size of the nucleus.

Figure 6.3 shows a hand and a table atom. It really does not matter which is which, as we shall see. Both atoms have a positively charged nucleus (red) in the center and negatively charged electrons (blue) in outer orbits far from the nucleus. Now that we are looking on the atomic level, we can consider what happens when we try to bring a hand in contact with the table. As we move the two atoms closer to each other, the electrons in one atom will begin to repel the electrons in the other atom. As we try to move them closer, this repulsive force will get stronger and stronger. Before the atoms can touch, the very large repulsive force will push them away from each other, so there is never any contact. It is the electric field of, say, the table electrons that is acting over some, albeit short, distance that is causing the repulsive force on the hand atoms. So, in fact, there are no contact forces, but only noncontact forces due to the fields. When you decide to get up and walk away, not only will the floor be pushing you, but your feet will be "floating" slightly above the floor, never really touching it.

Do you have a question here? The above discussion explains why there really is no contact between objects due to the repulsion of the atomic electrons. But I am hoping you are now asking how anything stays together. What holds this book together or the table or even you? Think about this. Hopefully it will lead to a good class discussion.

6.5 END-OF-CHAPTER GUIDE TO KEY IDEAS

- Can you describe the idea of action at a distance?
- Can you describe the field concept?
- How do you detect an electric field? How would you detect a gravitational field?
- What is the evidence that the field concept is correct?
- What is light?
- Can you explain why two objects are never in contact?

QUESTIONS/PROBLEMS

1. Write down any questions you still have about the field concept or action at a distance that you would like discussed further in class.
2. Define noncontact forces.
3. What are the two ideas that attempt to explain noncontact forces? Describe them in your own words.
4. What produces an electric field?
5. What produces a gravitational field?
6. What does the gravitational field act on to produce a gravitational force?
7. Given Equation 6.2, write the equivalent expression for the gravitational field.
8. In Equation 6.2, what is the meaning of the symbol d?
9. In action at a distance, what causes a force on an electric charge?
10. In the field concept, what causes a force on an electric charge?

11. What is the evidence that causes us to believe the field concept is correct?
12. What is necessary for light to be produced?
13. What is light?
14. Give a quantitative definition of *being in contact*.
15. Explain why the statement "you are really not in contact with the book that you are holding" is correct.
16. Choose which statements are true or false. For those that you choose as being false, explain why and correct them to make them true.
    a. In order to have an electric field, there must be at least two charges present.
    b. Light is produced when charged particles are moving.
    c. A single charged particle could create both an electric and magnetic field under the right circumstances.
    d. The electric force on a charged particle is produced by the action of the electric field acting on that particle.
    e. The source of a magnetic field is positive or negative charges that must be in motion.
    f. The source of the electric field is positive or negative charges whether they are in motion or at rest.
    g. The electric field acts on the charge that produces it.
    h. The electric field never acts on the charge that produces it.
    i. The electric force is caused by two electric fields acting on each other.
17. Given the discussion about why objects are not really in contact, explain why, for example, this book or you are held together.
18. Given the discussion about why objects are not really in contact, explain how a knife cuts through butter.
19. Is the field concept consistent with Newton's third law? Explain.

7 The Character of Natural Laws

We have already studied quite a few different but specific laws: the law of gravity, the laws of motion, and the laws of electricity and magnetism. What we will discuss in this chapter are some overriding principles that most, if not all, laws seem to follow. This not only tells us something deep about the nature of nature, but also demonstrates the unity of physics.

7.1 CAUSALITY

The principle of causality is really quite simple. It states:

The cause must always come before the effect.

At this point you are probably thinking to yourself that is pretty trivial. So what's the big deal?

Well, because it is so primary, it is an important test for any theory. If there were a theory proposed that violated causality, we would know that theory could not be correct. In fact, when Einstein introduced his theory of relativity, some people claimed it could not be correct since it seemed to violate causality. Relativity does indeed predict that different observers moving relative to each other would observe different events occurring at different times. For example, observer 1 could observe event A occurring before event B, while observer 2 could observe the same two events such that event B would occur before event A. At first glance this appears to violate causality. But causality only has to do with events that are causally related; i.e., one has to be the cause of the other (the effect). In that case, it can easily be shown that relativity predicts that all observers will see the events in the correct time order. But if this were not true, then we would have to reject relativity as a valid theory due to its violation of the principle of causality.

7.2 THE PRIME DIRECTIVE

I have borrowed this name from a popular TV series, *Star Trek*. The reason is that the statement of this principle sounds like a directive. It states:

If anything can happen, it will happen.

Or the stronger version:

If not forbidden, it must happen.

It tells us that in nature anything goes—unless. Unless what? Unless there is a law preventing it. If not forbidden by some law, any action, happening, event (however you want to designate it) will occur no matter how unlikely it seems. It turns out to be a very important principle, especially when combined with some of the ideas in quantum mechanics, as we shall see in a later chapter.

It is also very useful in helping us discover new laws or properties of nature. If at a given time, the laws we know allow for a certain event to occur, but that event is never seen, then it could be that this is a hint that there is a yet undiscovered law preventing the occurrence. This is indeed what happened with the discovery of a new property of matter called strangeness, along with a new law known as the conservation of strangeness. We will discuss this further in a later chapter about quarks and the basic building blocks of matter.

7.3 SYMMETRY

This is an idea with which I assume you are familiar. There are certainly many symmetric objects all around us—flowers, snowflakes, tile floors, globes, and even people, to name a few. The two shapes below are well-known symmetric figures.

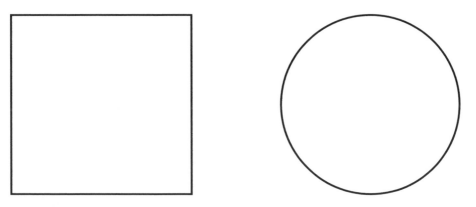

FIGURE 7.1   Two familiar and symmetric shapes.

What makes these things symmetric? Is one more symmetric than the other? How do you know? See if you can answer these questions and try to think of a general definition as to when something is symmetric.

What we really want to discuss in this section is not the symmetry of geometric objects, but the symmetry of the laws of nature. In order to even understand what it means for a law to be symmetric, we have to have a good definition of symmetry. It turns out that the mathematician Hermann Weyl (1885–1955) has given us a definition that pertains to both objects and laws. His definition states:

Something is symmetric if you can find something to do to it such that after you have done it, it still looks the same (you cannot tell you have done it).

Note, this says *something*, not *anything*.

Let us look at the square in Figure 7.1 to better understand this. What can be done to the square such that afterwards it looks the same? It turns out there are several things that can be done. It can be rotated through an angle of 90 degrees, or 180 or 270 degrees. If this were done, and you were not looking, there would be no way for you to know that anything was done to the square. If the square was alone (not with the circle), it could also be flipped or reflected, and it would still look the same. There are certainly lots of things that can be done to the square such that you could easily tell that something was done. For instance, it could be rotated through any angle that is not an integral multiple of 90 degrees, or it could be stretched or mangled in some way. But we were able to find *something* that could be done to leave it unchanged, and that is what makes the square symmetric.

Now let us look at the circle in Figure 7.1. What can we do to it such that it still looks the same? Hopefully you see there are many more things we can do. Instead of only three angles, it can be rotated through any angle and it still looks the same. It can also be flipped or reflected just as the square. In answering the question above about which was more symmetric between the two shapes, I suspect most of you chose the circle. We now see, using Weyl's definition, why that is. There are more things we can do to the circle that leaves it unchanged compared to the number of things we can do to the square. So, Weyl has done us a double service. He not only has given us a very nice definition of symmetry, but he also has given us a way of measuring the degree of symmetry.

Two clarifying remarks should be made here. First, the phrase "something to do to it" in the definition is a bit vague (fuzzy?). A better phrase is "that one can perform an operation" such that after…. For instance, the operations we mentioned above are rotation and reflection. Also, if something does not change, it does not vary, or is left *invariant*. We will see in the next chapter why *invariant* is such a nice word. So, let us rewrite Weyl's definition using our new terminology.

Something is symmetric if we can find some *operation* such that afterwards it is left *invariant*.

7.4   SYMMETRY AND THE LAWS OF NATURE

The idea that the laws of nature must have certain symmetries associated with them has become an important and central theme across many fields of physics. It has led us to new insights as to how nature works. Weyl's definition gives us a way of understanding how laws can be symmetric, and why it is so important. The symmetry of the laws can be expressed either mathematically or experimentally. In fact, they are just equivalent ways of showing the same thing.

7.4.1 Space Translation Symmetry

We have actually used the idea of this symmetry way back in Chapter 2 when we discussed the assumptions intrinsic in the definition of physics. One of the assumptions was that the laws of physics were the same everywhere in the universe. In other words, we can change our location in space (i.e., translate our position from one location to another) and the laws will not change. The operation that was performed was that of translation. Figure 7.2 demonstrates this and will also help us to see how this works mathematically.

FIGURE 7.2   Two masses, M_1 and M_2, separated by distance d. The right-hand combination shows the same two masses, but translated in space by a distance X.

Consider the gravitational force between the two masses M_1 and M_2. Newton's law of gravity is

$$F = GM_1M_2/d^2$$

This tells us that the force depends on the distance d between the two masses. As we see in Figure 7.2, the force on the left or right must be the same even though the two masses have been translated in space by the distance X. Mathematically we can easily see this since in the left-hand case $d = (x_2 - x_1)$, while for the right-hand masses, $d = (X + x_2 - (X + x_1)) = (x_2 - x_1)$. So the value of d remains invariant even though the positions have changed by the amount X. Thus, Newton's law of gravity is invariant under the operation of translation in space. So is Coulomb's law of electric force since it too depends only on the distance d. In fact, we believe all laws of nature must be symmetric under space translation.

This, then, sets a very strong constraint on the mathematical form of any law. Namely, a law must depend on the *difference* between two positions. It could not depend, for instance, on the position itself or on the sum of two positions. In both these cases such a law would give a different prediction depending on the location of an object in space. We would know that such a proposed law could not be correct.

A more dramatic way of looking at this is to state that the laws of nature cannot depend on where a particular physicist decides to choose the origin of the coordinate system. All numerical values of position depend on this choice. For instance, let us say you are doing an experiment on a table. One person decides to measure all positions from the edge of the table, while another experimenter decides to measure from the center of the table. Each person would get a different value for a particular position on the table. But the predicted results (and the actual results of the experiment) cannot depend on this choice of origin. By having only differences in position enter into the equations, this is insured.

There are two other space symmetries that also tell us important things about the nature of nature. They are the symmetries of space rotation and space reflection. We will discuss the latter below.

7.4.2 TIME TRANSLATION SYMMETRY

This is similar to space translation symmetry, but it deals with translation in time instead of space. It says that the laws of nature cannot depend on an instant of time. In other words, the result of an experiment cannot depend on the time it is performed. We can again see how this works mathematically by considering Newton's second law of motion. For convenience, let us write it as $F = ma$:

$$F = ma = m\Delta v/\Delta t = m(v_2 - v_1)/(t_2 - t_1)$$

We see that the time comes in as a time *difference*, so if we were to shift all times by an amount T (i.e., do the experiment at a different time or measure all times from a different starting time, for instance, from the time of the big bang some 13 billion years ago), this time T would cancel out in the difference. Just as in the case of space translation symmetry, this also puts constraints on how time must come into the equations of physics. The mathematical forms of the equations cannot depend on the absolute time or the sum of times, only on time differences.

7.4.3 TIME REVERSAL (REFLECTION) SYMMETRY

This symmetry says that nature should behave the same whether time is running forward or backward. In other words, if a motion picture was taken of some events, you would not be able to tell which way the projector was being run, forward or backward. At this point I suspect you are all saying that this is nonsense since there are many situations in which it is perfectly clear which way the projector is running. To give just one example, consider dropping a glass that shatters into thousands of pieces upon breaking. The backward-running projector would show thousands of separate pieces all coming together to form a perfect glass that would then leap into the air—obviously not possible.

Indeed, on the macroscopic scale time reversal symmetry does not hold, but on the microscopic scale it does. On this scale we are considering atoms that look like little balls that are in some sort of motion. If we could take a picture of atoms and run it backwards, we would see atomic balls in perfectly reasonable motion. So, we could not tell which way the projector is running.

I mention this particular symmetry because it is one of the rare cases where we cannot predict the macroscopic behavior of matter from the microscopic, atomic behavior. The fact that macroscopic time reversal is not a good symmetry has important consequences for our society, especially as to what is sometimes referred to as the energy crisis. This has to do with concepts such as entropy and the second law of thermodynamics. We will discuss this further in Chapter 8. Some of you may have heard something about it already.

7.4.4 MATTER-ANTIMATTER SYMMETRY (MATTER REFLECTION)

Some of you may have heard about antimatter before, possibly in some science fiction story. Antimatter is real but very rare. It can be produced in particle accelerators, in some radioactive decays, and naturally in cosmic rays from outer space. In fact, for every type of ordinary matter there is its corresponding antimatter. An antiparticle has the same mass as its particle partner, but all other properties are opposite. For instance, the electron has a negative electric charge, while its antiparticle, known as the positron, has a positive charge. Similarly, the antiproton has a negative charge compared to the proton, which is positively charged.

Charge and mass are not the only properties particles have. If you remember, in the discussion about how iron can be magnetic in Chapter 4, it was mentioned that electrons spin around an axis similar to a top. This is another property of elementary particles, and antiparticles spin in the opposite direction. In fact, the neutron has no electric charge, but the antineutron does indeed exist, and it spins in the opposite direction as the neutron.

You may also have heard that when matter and antimatter meet, it is quite violent. Both masses disappear and turn into pure energy according to Einstein's famous equation, $E = mc^2$, which tells us how much energy is released. This is known as annihilation. The matter and antimatter annihilate each other into pure energy.

Matter-antimatter symmetry implies that matter and antimatter should behave identically. As one example, an antihydrogen atom made up of a positron and an antiproton should have the exact same properties as an ordinary hydrogen atom. As far as we know, this is true. Another way of looking at this, using Weyl's definition, is that if we overnight changed every particle in the universe into its antiparticle, when we woke up in the morning, there would be no way of telling that this change had occurred.

> Do you have a question here? Hopefully, what you have just read in the paragraphs above has disturbed you a bit and has evoked a question in your mind. If it has not, think about it before reading on.

It was stated above that matter and antimatter behave identically. But it was also stated that antimatter is very rare. If they behave identically, shouldn't there be equal amounts of each? It turns out that this is a very profound question.

We have very good experimental evidence that antimatter is not just rare on earth, but throughout the entire universe. At the same time, we are almost positive that at the big bang, the beginning of the universe, there had to be equal amounts of matter and antimatter—in other words, they were indeed symmetric. So, what happened? We do not know the details, but we believe that sometime in the very early universe, within a fraction of a second after the big bang, the symmetry between matter and antimatter had to be broken. While, as was said, we do not know exactly what happened or how, we do know that the size of the asymmetry was extremely small, about one part in a billion ($1:10^9$).

How do we know this? Remember that above we talked about the fact that matter and antimatter annihilate each other when they meet. If there was perfect symmetry between the two, then there would have been just enough antimatter to annihilate all the matter, and there would be neither matter nor antimatter left in the universe, but just pure energy in the form of electromagnetic radiation (light). Another way of saying this is that the ratio between matter and light energy would be zero. But, of course, it is not. We have measured this ratio to be $1:10^9$. This must be the size of the asymmetry that occurred in the early universe.

At this point you may be saying to yourself that this is all somewhat interesting, but why should you really care? Well, you should care very much. If this symmetry were not broken, then there would be no matter in the universe at all. The sun and the earth could not exist and neither could you. So, let us be very thankful for this broken symmetry. In fact, while we have seen that the idea of symmetry is extremely important, we have learned that the idea of broken symmetries is also important. The matter-antimatter broken symmetry is not the only one, and we owe our existence to some of the others also.

7.4.5   SPACE REFLECTION SYMMETRY (PARITY)

To be complete, we should mention the third of the symmetries: reflection. In fact, space reflection does have to do with what we ordinarily think about when we hear the word *reflection*, i.e., reflection in a mirror. If this is good symmetry, then it says we could not tell whether we were looking at some action directly or viewing it in a mirror.

At this point you might be thinking, "Of course I could tell, if someone was wearing a t-shirt with writing on it. The writing would look backwards in the mirror image." But what this symmetry states is that you could tell only if what appeared in the mirror image was *physically impossible*. Backward writing is not impossible. In fact, you can go to your friendly t-shirt store and have the writing on the t-shirt printed backward.

Until 1958 all physicists assumed space reflection (parity) was indeed a good symmetry. But in that year, two physicists, Lee and Yang, suggested that in processes where the weak nuclear force acted, parity was not a good symmetry. They were correct, as demonstrated by an experiment done shortly after their prediction. For that, they were awarded the Nobel Prize.

We will not go into any further detail here except to note the interesting fact that in one way or another, the three reflection symmetries seem to be broken. On the other hand, as far as we know, the translation symmetries are not.

7.5   END-OF-CHAPTER GUIDE TO KEY IDEAS

- What is causality? What is its significance?
- What is the meaning of the prime directive? What is its significance?
- Define *symmetry*.
- What are the "translation" symmetries?
- What are the "reflection" symmetries?
- In what ways are the three reflection symmetries broken?

QUESTIONS/PROBLEMS

1. What is the significance of this chapter?
2. What are the three characteristics of natural law that are discussed in this chapter? Define each of them.
3. What are the two versions of the prime directive?
4. How do you know if an object is symmetric?
5. What is the criterion for one object to be more symmetric than another?
6. What is the single word that can be used to describe something that does not change?
7. What are two equivalent ways in which the laws of nature are symmetric?
8. List the "translation" symmetries. What is the definition of each?
9. List the "reflection" symmetries. What is the definition of each?
10. Show mathematically that Coulomb's law is invariant under space translation symmetry.
11. Is time reversal a good symmetry in all cases? If not, give an example when it does not hold. Give an example when it is a good symmetry.
12. What is antimatter. How does it differ from ordinary matter?
13. What is annihilation?
14. How do we know the matter-antimatter symmetry has been broken? By how much has it been broken? Why is this important to us?
15. In the list below are examples of possible laws of nature. In each case, tell whether it could be a good law or would violate space translation symmetry, time translation symmetry, or both. (x and x' represent positions in space; t and t' represent instants of time.)
    a. $F = Kx$
    b. $F = K(x + x')$
    c. $F = K(x - x')$
    d. $F = Kt$
    e. $F = K(t + t')$
    f. $F = K(t - t')$
    g. $F = Kx(t - t')$
    h. $F = Kxt$
    i. $F = K(x - x')/(t - t')$

8 Conservation Laws

Conservation laws are extremely important in physics. They are relevant to a wide variety of physical phenomena. They are very powerful and hence very useful. And interestingly, they follow directly from the symmetry principles we discussed in the last chapter. We can see this by just considering what it means if something is conserved.

If something is conserved, it does not increase or decrease; it does not change; it does not vary. In other words, it remains *invariant*. So, we can give the following definition to conservation:

Something is conserved if we can do something to it such that afterwards it remains invariant.

Of course, this sounds almost identical to our definition of symmetry. It is no accident. Emmy Noether, a woman mathematician, proved that symmetry principles lead directly to conservation laws. For instance, space translation symmetry leads to the conservation of momentum, and time translation symmetry leads to the conservation of energy. We will discuss both of these conservation laws in this chapter. Using these as examples, we will be able to see why conservation laws are so powerful.

8.1 CONSERVATION OF MOMENTUM

The word *momentum* is often used in everyday speech to imply that something is in motion. But in physics, momentum has a very well-defined meaning. The momentum of a single object with a mass, m, is given by the equation

$$\mathbf{p} = m\mathbf{v} \tag{8.1}$$

Momentum is a vector quantity since it is directly proportional to the velocity vector. The units of momentum are [Kg][M]/[s]. Many quantities in physics have units named after some famous physicist. For instance, the unit of force is the Newton for Isaac Newton. But no one has been honored for momentum.

If there is a group of objects and we are interested in the momentum of the entire system, then we just add up the momentum of the individual objects to get the momentum of the system. We will designate the momentum of a system of objects by **P**, where we can write

$$\mathbf{P} = \Sigma\mathbf{p} = \mathbf{p}_1 + \mathbf{p}_2 + \mathbf{p}_3 + \dots \tag{8.2}$$

Do not worry, for almost anything we will be doing, there will be at most two objects in the system.

We are now ready to state the conservation of momentum:

The momentum of a system is conserved *if* there are no external forces acting on that system.

Notice there is an *if* in the statement of the law. So, momentum is not always conserved. It depends on the condition of whether there is an external force acting on the system. How do we know? Well, it depends on what the system is, since *external force* means a force applied from outside the system. Well, then, how do we know what the system is? Here is the nice part—you are free to choose the system to be whatever you like. Let us see how this works with a specific example.

THE BALL IN THE AIR

The Ball Is the System

Consider throwing a ball up into the air and we will choose the system as the ball. Let us consider the ball only while it is in the air (after it has been thrown and before it has been caught). Is the momentum of the ball conserved while it is in the air? There are several ways of analyzing this. Think about this before reading further.

Some of you may remember learning that when something is thrown into the air, at the same height, it has the same speed going up as going down. That is true if we ignore the very small effect of air friction, which we will almost always do. Thus, you might have answered yes, the momentum of the ball is conserved. But, remember momentum is a *vector* and the velocity vector has the opposite direction on the way up compared to the way down. So, in fact, the momentum of the ball is not conserved. We can also see this in two other ways. First, the speed, and hence the velocity, is changing throughout the motion, slowing down on the way up and speeding up on the way down. Momentum is not conserved at any time during the motion. Second, according to the statement of the law, momentum is not conserved if there is an external force on the system. In this case, there is indeed an external force on the ball (the system in this case). It is the gravity of the earth. So, we see in the case of having an external force on the system, the momentum has not been conserved.

A Bigger System

Now, let us consider the ball again but this time as part of a system. Choose the system as being made up of the ball, the person who throws the ball, and the earth. We will consider the motion from the time before the ball is thrown to any other time during its motion. We will see the momentum of this system is, in fact, conserved. For the purposes of this example, we will (and can) ignore the external effects of the sun and moon. They have minute effects on the motion of our system over the time of a few seconds that the ball is in flight. Thus, there are no external forces on the chosen system, and we expect the momentum of this system to be conserved. To check this out, let us consider the momentum of the separate parts of the system. For our purposes here, the thrower of the ball can be considered as part of the earth.

When something is conserved, it does not change. This can be expressed as an equation in one of two ways:

$$\Delta P = 0 \tag{8.3}$$

or

$$P_2 = P_1 \tag{8.4}$$

where the subscripts 1 and 2 refer to two different times t_1 and t_2. In words, this just says that if momentum is conserved, then the momentum at one time must be the same as (equal to) the momentum at any other time. Since we know the momentum of the system before the ball is thrown must be zero, then it must always be zero. To see where this leads, consider t_2 as being sometime while the ball is in the air. In addition, write P_2 as the sum of the momenta of its individual parts. This leads to

$$P_2 = p_{ball} + p_{earth} = 0 \tag{8.5}$$

or

$$p_{ball} = - p_{earth}$$

This tells us that the momentum of the earth must always be equal to the momentum of the ball (of course, in this example only). The $-$ indicates that the earth and ball are moving in opposite directions. At this point, you may be wondering how the earth got any momentum at all. Before answering this, let us first see what the relation between the speed of the earth and ball is.

Remembering that the definition of momentum is $p = mv$, and just concentrating on magnitudes, we can write

$$M_{earth} v_{earth} = m_{ball} v_{ball} \tag{8.6}$$

or

$$v_{earth} = (m_{ball}/M_{earth}) v_{ball}$$

Since the earth has a mass of about 10^{24} kg and a ball's mass is less than a kilogram, the ratio (m/M) is minute. This tells us that while the earth's momentum is equal to that of the ball's momentum, the speed of the earth is immeasurably small. That is why no one feels the earth moving whenever a ball is thrown.

Note that the relation between the momentum of the earth and the momentum of the ball being equal and opposite is true at any time during the motion. So, when the ball is moving upward, the earth is moving downward, and when the ball is moving downward, the earth is moving upward. How does the earth know what to do? This goes back to the question about how does the earth acquire any momentum at all.

While it is true that there are no *external* forces acting on the system, there are several *internal* forces acting on different parts of the system. But all of these forces form action-reaction pairs (remember Newton's third law), so their net effect on the system is zero since they must all cancel out. For instance, in order to throw the ball into the air, the person who throws it must apply an upward force on the ball. The ball, by the third law, applies an equal and opposite force on the person's hand. This downward force gets transmitted to the earth via the bones in the person's body. Thus, the earth receives a downward force, causing it to accelerate and hence acquire momentum. Once the ball is in the air, there is no contact between ball and earth, but there is still the gravitational attraction between them. As the ball goes up, it slows due to the earth's gravity pulling it down. For the same reason, the downward motion of the earth slows due to the ball's gravity pulling it up. Remember, by either considering the third law again or just looking at the law of gravity, the gravitational force on the ball due to the earth is equal in magnitude to the gravitational force on the earth due to the ball. As the ball reaches its highest point and stops, the earth reaches its lowest point and also stops. Then, as the ball goes down, the earth moves up. Finally, when they come in contact again, the equal and opposite collision forces between them bring them both to rest. The momentum of the earth-ball system is zero at all times.

It should be noted that since the idea of momentum has to do with motion, it should not be surprising that Newton's laws of motion are important in understanding the conservation of momentum. First, the second law is $a = F_{ext}/m$, which tells us if F_{ext} is zero, acceleration is zero. But if acceleration is zero, then the velocity, and hence momentum, does not change. Also, in a system of two or more objects, there can certainly be forces between those objects, but due to the third law, those forces within that object must all cancel. Thus, internal forces cannot affect the motion of any system. They can affect the motion of parts of the system, but not the system as a whole.

AN EXPLOSIVE EXAMPLE

After reading the above example, you may be asking that if conservation of momentum is just a consequence of Newton's second and third laws, why go to the trouble of introducing the idea of momentum, a new variable and concept that does not seem to be needed. The following example will answer that question.

Consider a bomb of mass $m_b = 10$ kg that is initially at rest. At some instant it explodes into two parts, one of which has a mass of 6 kg and a speed of 10 m/s. The problem is to find the speed of the second mass. To help us see what is happening, consider Figure 8.1.

First, notice that in the above figure, we have already applied a conservation law: the conservation of mass. That is how we knew that the second piece had a mass of 4 kg. Now, can we apply conservation of momentum? Yes, if our system has no external forces acting on it. In this case, our system is the bomb (both before and after the explosion), and since the explosion is internal to the bomb, we can indeed invoke the conservation of momentum. In fact, it was already used in Figure 8.1. With the momentum of the system being zero, then the two pieces of the explosion must fly off back-to-back in order for their individual momenta to cancel.

In order to use the principle to get the answer we are looking for, we write the equation stating the conservation of momentum as before (Equation 8.4):

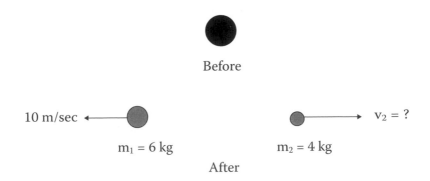

Before

10 m/sec $\leftarrow$ $m_1 = 6$ kg $m_2 = 4$ kg $\rightarrow$ $v_2 = ?$

After

FIGURE 8.1 **(See color insert following page 102.)** Explosion of a 10 kg bomb initially at rest.

$$\mathbf{P}_2 = \mathbf{P}_1$$

In fact, anytime you want to, and are able to, use momentum conservation, the above equation is the starting point.

Since, as before, the intial momentum was zero, we have $\mathbf{P}_1 = 0$. In order to find the speed of the second part, we are interested in a time, t_2 after the explosion, so we can write

$$\mathbf{p}_1 + \mathbf{p}_2 = \mathbf{m}_1\mathbf{v}_1 + \mathbf{m}_2\mathbf{v}_2 = 0$$

where now the subscripts 1 and 2 refer to the two parts of the bomb after the explosion. In order to get rid of the vectors, which we do not want to deal with, we just have to define directions. We can do this somewhat arbitrarily as long as we are consistent. The convention we will adopt is that anything going to the left, in Figure 8.1, has a negative velocity, and anything to the right has a positive velocity. With that convention, we can write

$$m_1(-v_1) + m_2(v_2) = 6(-10) + 4(v_2) = 0$$

where we have put in the given numbers where appropriate. This is now an easy algebraic problem to solve for v_2:

$$4v_2 = -6(-10) = +60$$

or

$$v_2 = 60/4 = 15 \text{ m/s}$$

Our answer is that after the explosion the second piece flies off with a speed of 15 m/s toward the right. While this answer might look rather mundane, it is really quite a remarkable result. Note that an explosion is a rather violent and complicated event. Was it mechanical, or chemical, or maybe even nuclear? It was never mentioned, and yet we could calculate the desired answer. This is the real power of the conservation of momentum and, for that matter, just about all conservation laws. If something is conserved, then it does not matter what the complicated details of the interactions might be, the quantity is still conserved. So, we can be completely ignorant of gory, messy details, which are many times unknowable, and still be able to calculate desired results. We will see how this also works for the conservation of energy, which will be discussed in the next section. But before discussing energy, let us consider one more example of the conservation of momentum.

In the two examples we have considered above, the initial momentum, in each case was zero. Just to make sure we see that this does not have to be so, let us consider a case when the initial momentum is not zero. Let us use the bomb example above, but with the bomb having an intial speed before it explodes. In addition, we will have the piece with the known speed moving in the same direction as the initial velocity of the bomb. We will see the purpose of this choice below. Figure 8.2 shows this situation.

We do this problem in exactly the same way as we did the previous one, only the numbers are different. So, we first write our conservation of momentum equation:

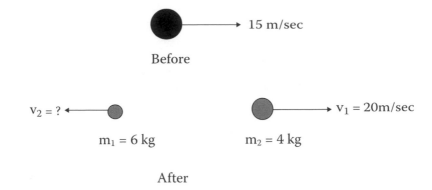

Before

After

FIGURE 8.2 (**See color insert following page 102.**) A 10 kg bomb moving with a speed of 15 m/s before exploding.

$$P_2 = P_1$$

This time the initial momentum is not zero, but has the value of $m_b v_b = 10*15 = 150$ kg-m/s. We therefore write

$$m_1 v_1 + m_2 v_2 = 6(20) + 4(-v_2) = 150$$

or

$$-4v_2 = 150 - 120 = 30$$

or

$$v_2 = -7.5 \text{ m/s}$$

What does the minus sign mean in the answer? It means that our assumption as to the velocity direction of the 4 kg mass was incorrect, and that it too moves to the right. Note that in Figure 8.2, we have chosen (assumed, guessed) that the 4 kg mass went to the left, which means, according to our convention, that in the equation above, we must assign its velocty as being negative $(-v_2)$. The minus sign in the final answer indicates that this guess was wrong. Note that if the bomb was not moving initially, then the two pieces would have to fly off back-to-back. But if the bomb is moving initially, then the second piece could go off in either direction, depending on the details of the problem.

That is another powerful thing about the conservation of momentum: it automatically corrects wrong guesses. But how can the two objects both be flying off to the right? Because conservation of momentum requires it. Let us see how. The initial momentum was 150 kg-m/s to the right. Thus, after the explosion, the system must still have this same momentum. But the 6 kg piece only has a momentum of 120 kg-m/s. So the 4 kg piece must therefore have a momentum of 30 kg-m/s, also to the right, to conserve the momentum of the system.

8.2   CONSERVATION OF ENERGY

Energy is one of the most important quantities in modern society. Most of us probably use the word several times a day without even realizing it. Our homes need energy. Our cars need energy. We need energy. Just about every aspect of our industrialized world requires energy, and because of this, we even have concerns about an energy crisis. So, it is important to understand what energy is and why it is so useful.

Energy is defined as the ability to do work. *Work* is a commonly used word, but in physics it has a very definite meaning. Work is done when a force, F, moves something over a distance, d. It can be calculated by the equation W = Fd. This is different than the common usage when we might feel we are doing a great deal of work, for example, when we hold a heavy object in the air. We will probably get pretty tired if we hold it for a long time, but from a physics point of view, no work is being done. But if something has the *ability* to do work, then it has energy. Because this ability can manifest itself in different ways, energy comes in different forms. One form can be transformed into another form. It is this property that makes energy so useful.

One important aspect of energy is that it is a scalar quantity. In other words, it does not depend on direction as momentum does. Also, energy must have the same units as work. From the equation for work above, this is Newton-meters or kg-m^2/s^2. In the MKS system, the unit of energy or work is called the Joule.

The conservation of energy simply stated is: energy is always conserved. The equation representing this can again be written in two ways:

$$\Delta E = 0 \qquad (8.7)$$

or

$$E_2 = E_1 \qquad (8.8)$$

Note that there is no *if* here, as there was for conservation of momentum. We will have an example of using the conservation of energy below. But before we do, we must first discuss the different forms of energy.

8.2.1   THE DIFFERENT FORMS OF ENERGY

1. Kinetic energy: This is probably the most familiar form. It is the energy of motion. The equation for kinetic energy is

$$KE = (1/2)mv^2 \qquad (8.9)$$

   If something is moving, it has the ability to do work due to the fact that it can hit something, producing a force due to the collision that can move the struck object over some distsnce. For instance, a moving hammer does work on a nail when it drives it into a piece of wood.

2. Potential energy: Some of you probably have learned something about this form of energy, but it is easily misunderstood. It is an energy having to do with

the position of an object. It does not matter whether the object is in motion or not. In order for an object to have potential energy, it is necessary that the object is subject to a certain type of force that has a special property. Namely, the force acts in such a way to either bring the object to a definite location or repel the object from a definite location. Such a force is called a conservative force. Gravitational, electrical, and spring forces are such conservative forces. For instance, the gravitational force due to the earth acts on every massive object in a direction toward the center of the earth. A spring, whether stretched or compressed, acts to bring an attached object back to the unstreched position. Friction is a good example of a nonconservative force. It always acts opposite of the direction an object is moving, not toward a definite location.

An object has a potential energy when it is under the influence of a conservative force but is not at that special location associated with the particular conservative force. Since the force tends to act to bring it to that location, it would accelerate the object, causing it to gain speed and hence kinetic energy. Even if the object cannot move due to some other force acting on it in an opposing direction, it still has the potential to be able to move—hence the name *potential energy*. For instance, if you are holding a ball in your hand above the floor, the ball cannot move because your hand is preventing it from falling, but gravity is still acting on it. It has potential energy because if you released it, it would then be able to fall.

For potential energy, there is not just one equation like kinetic energy. This is because there are different expressions, depending on the particular conservative force. The only one we will discuss here is the gravitational potential energy near the surface of the earth. It is given by the expression

$$PE = mgh \qquad\qquad (8.10)$$

where m is the mass of the object, g is the acceleration due to gravity, and h is the height of the object measured from the lowest point the object can reach. Some of you may remember that (mg) is the weight of an object, so the gravitational potential energy an object has near the earth is just its weight (the gravitational force due to the earth) multiplied by its height.

We will combine the gravitational potential energy with kinetic energy to help us understand why the conservation of energy is so powerful and useful. But before we do this, let us list and briefly discuss the other forms of energy.

3. Heat energy: This is the random energy of motion of the molecules in a substance. So, it is really just a special form of kinetic energy.

4. Atomic energy: Usually this is confused with nuclear energy. Technically, atomic energy is the energy associated with the atom. Sometimes it is known as chemical energy. Since, as we have discussed before, the atom is held together by the elctrical force of attraction between the positively charged nucleus and the negative electrons, atomic energy is really electrical potential energy plus the kinetic energy of the electrons in their atomic

orbits. One reason for the confusion has to do with the fact that the first nuclear weapon was mistakenly called the atomic bomb.

5. Electromagnetic energy: This comes in different forms, but because all forms have to do with charges either at rest or in motion, we will lump them all together. For instance, there is electrical potential energy, as we have mentioned above. There is energy stored in the electric and magnetic fields. Finally, light (electromagnetic radiation) has energy associated with it. Think about how hot you feel on a bright sunny day.

6. Nuclear energy: This is due to the forces between the neutrons and protons in the nucleus of the atom. It is nuclear potential energy.

7. Mass energy: This is truly an independent form of energy. It is through the famous eqaution $E = mc^2$ that Einstein showed that mass and energy are equivalent. Mass can be converted to energy, and energy can be converted to mass. We will discuss this further in later chapters.

8.2.2 CONVERSION OF ENERGY

Why is energy such an important quantity in our industrialized world? Basically because it does come in different forms, and one form can be transformed into another. For instance, waterfalls such as Niagra Falls are used to generate hydroelectric power. This is because the gravitational potential energy of the water at the top of the falls is converted to kinetic energy as the water falls. The falling water strikes a turbine, causing the turbine to get rotational kinetic energy. The turbine is rotating in a magnetic field, which creates electrical energy due to Faraday's law. The electric energy, in the form of electric current, finally gets to our houses or offices, where it is then converted to heat, light, or other forms of energy.

By the way, the only difference between a hydroelectric generating plant and a coal or nuclear plant is the source of energy to turn the turbine. In a coal plant, coal is burned (release of chemical energy) to heat water and create steam (fast-moving molecules with kinetic energy) to force the turbine to turn. In a nuclear plant, nuclear energy is released in a process called fission (splitting of atoms) to heat the water to make steam. If there is a steady wind, then the kinetic energy of the air molecules can cause a turbine to rotate. The only form of electrical energy conversion that does not finally rely on Faraday's law is solar conversion. Here, the energy in sunlight causes electrons to be emitted from a metal surface, producing an electric current. This process is known as the photoelectric effect. It is the same principle that is used in a light meter in a camera.

8.2.3 A SPECIFIC EXAMPLE: THE ROLLER COASTER

Have you taken a ride on a roller coaster? If you have, you know it is a thrilling and chilling experience. We will use a roller coaster as an example of the use of the conservation of energy.

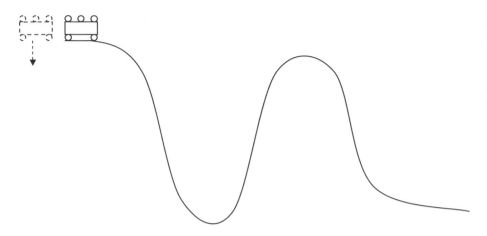

FIGURE 8.3 A roller coaster track and car. The thrills begin when the car is released to follow the track at a height h above the ground.

Basically, a roller coaster is a car that is forced to travel on a rather complicated, hilly track (see Figure 8.3). The car is first pulled up to the highest point and is then released to freely move along the track. The thrills and chills are due to the large speeds and accelerations experienced during the ride along the track. An important quantity the designer of a roller coaster has to know is the final speed the car has at the bottom of the track. He or she has to know this in order to be able to design the braking system and length of track necessary to stop the car. It might appear that this is a very difficult task since you would have to know the exact shape of the track in order to know all the forces acting on the car at each instant. Once the force is known at each point, the acceleration can be computed (do not forget a = F/m). Each of these accelerations must be added vectorially (the direction is different at each point) to get the final speed. A good physics student with a computer could probably work this out in a few days. We will now see how much easier this is using the conservation of energy.

The energies involved in the roller coaster are kinetic energy and gravitational potential energy. We will ignore any heat energy since a good roller coaster is built to minimize this. Since energy is always conserved, we have the energy at the top of the track equal to the energy at the bottom. Thus, we can write:

$$(KE + PE)_{bot} = (KE + PE)_{top} \tag{8.11}$$

or

$$(1/2mv^2 + mgh)_{bot} = (1/2mv^2 + mgh)_{top} \tag{8.12}$$

where m is the mass of the car, including the riders. We can now put in some values that we know: $h_{bot} = 0$ since it is the lowest point in the motion and $v_{top} = 0$ since the car starts off at the top with a very small speed. Also, h_{top} is just h, the height above the ground. We can then rewrite Equation 8.12 as

$$1/2mv^2_{bot} = mgh$$

or

$$v_{bot} = \sqrt{2gh}$$

This is our answer, which took us a few lines of simple algebra. Note that the final answer does not depend on the shape of the track or the mass of the car, m. The mass cancels out since it appears in both the kinetic and potential energy.

We can also use our roller coaster example to highlight what we might call a physics-biology problem. Let us imagine that the roller coaster operator made a mistake such that at the top of the track he backed up off the track (the dashed car in Figure 8.3), causing the car to drop straight down to the ground. A physicist in the car decides he better quickly calculate the speed at which the car will strike the ground. Since our answer before did not depend on the shape of the track, we can see that he should get the same answer as we did above. We could also start over, using Equations 8.11 and 8.12, which would lead to the identical result.

A question: Why, in the case of going along the track, do the passengers safely arrive at the ground, while in the case of falling straight down they are all killed? After all, in both cases the speed at the ground is the same. Think about this before reading further.

The solution: In the former case, when the car reaches the ground, brakes are applied, gently bringing the car to a stop. In other words, the acceleration on the car and passengers is small. In the latter case, the car stops abruptly, causing a very large acceleration. Writing Newton's second law in the form $F = ma$, we see that a large force is necessary to sustain a large acceleration. If that force is too large, as is the case when the car strikes the ground from the direct fall, the bones in our body cannot supply that force and they break, causing all sorts of bad things in us.

In either case, the important point to emphasize here is that by use of the conservation of energy we were able to easily find the final speed, independent of the very complicated (or simple) shape of the track, in much the same way we were able to apply conservation of momentum to find the speed of part of the bomb independent of the complications of the explosion. Hopefully you see why conservation laws can be so powerful. In this chapter, we have only discussed two. In later chapters, we will learn about others.

8.3 A NONCONSERVATION LAW: THE SECOND LAW OF THERMODYNAMICS

We have just finished discussing energy and the fact that it is always conserved. This may have raised a question in your mind: If energy is always conserved, then why is it we hear about an energy crisis where we are concerned about running out of renewable energy sources? The key word here is *renewable*, or another word is *reusable.* While energy is always conserved, even as we change it from one form to another, it tends to be converted from a more to a less usable form. In fact, it takes more energy to put it back into a more usable form. Hence, it is not worth doing. Why this is so has to do with what is known as the second law of thermodynamics.

Thermodynamics, as the name implies, is concerned with systems where heat energy can flow into or out of the system. In fact, the first law of thermodynamics is just a restatement of the conservation of energy when there is the possibility of heat energy being exchanged. This, in fact, covers most processes in nature. The second law is one of the most interesting laws in physics. It is a nonconservation law.

The second law has to do with a quantity we have not discussed so far, known as the *entropy* of the system. Entropy is a measure of the randomness or disorder of the system. The more disordered a system is, the higher the entropy. The second law of thermoynamics states:

The entropy of a system increases or, at best, remains the same.
It never decreases.

We can express this statement as an equation. Using S as the symbol for entropy, we can write

$$\Delta S \geq 0 \tag{8.13}$$

To contrast this with the conservation of energy, we write equation 8.7 again:

$$\Delta E = 0$$

Equation 8.7 says the change in energy is zero, or the energy remains constant, which is another way of saying it is conserved. On the other hand, Equation 8.13 says that the change of entropy is positive or, at best, zero. In fact, for the vast majority of processes, the entropy increases.

At this point, you may have several questions, such as: What is the meaning of the quantity entropy? What are the consequences of the fact that it never decreases? And how does this have anything to do with an energy crisis?

In order to best understand entropy, let us consider several different processes and see what they have in common:

- Some blue ink is dropped into a glass of pure water. After a while the ink spreads out throughout the water, which eventually turns a shade of light blue.

- Papers are stacked neatly on a desk. A wind enters the room through an open window and blows the papers around the room.

- A lump of coal is burned in some sort of power plant. Afterwards the mass of that lump is still the same, but is now distributed as coal dust over a large area.

- A piece of ice is placed in some warm water. After a while the ice melts and the average temperaure of the water has decreased.

> Do you see a connection between these? Think about this before reading
> further.

In each case, the sytem has gone from a more ordered to a less ordered (more disordered) state. Another way of saying this is that the system has gone from a less random to a more randomized state.

In the first case, the blue ink was in a small, confined volume. But afterwards, it was distributed randomly throughout the entire volume of the water. The papers are certainly less ordered or more randomized after being blown around the room compared to being neatly stacked on the desk. Again, the coal as dust is much more randomized compared to when it is in a small, well-ordered lump. As for the ice, initially it is in a well-defined ordered volume with a well-defined low temperature. Afterwards, the ice has turned to water, and those water molecules, which were initially ice, are randomly distributed throughout the much larger volume of the water. Also, in this case, heat had to flow from the warmer water to the colder ice in order to melt the ice.

In all these cases the entropy has increased as the systems have gone from more order to more disorder. From this, it is natural to suspect that entropy is a measure of disorder. In fact, this is true. But now you may ask: Why does nature like to go to more disorder? The answer is quite logical, once you think about it a bit. The most disordered state is the most probable state. Entropy is also a measure of the probablity of the state of a system. Looking at it from a probabilty point of view, the second law of themodynamics makes sense. Nature wants to go to the most probable state. Since entropy is a measure of the probability, then the entropy in any process must increase since it is natural to go from low probability to high probability.

Let us discuss two examples to show that high disorder (randomness) and high probability go together. First, consider tossing ten coins and looking at how many are tails and how many are heads. The most random state is when half the coins are heads and half tails. The probability of this happening is 0.246 (almost 1 in 4 times). On the other hand, the most ordered state is when the coins are all heads, for instance. The probability of this happening is 9.8×10^{-4} (1 in 1,024). The most random state is 250 times more probable than the most ordered state.

For the second example, consider playing poker and being dealt five cards. The worst hand is when you get a random distribution without even a pair. The probabilty of this is 1 in 2 (50%). Contrast this to the best hand you can get, a royal flush. This is a highly ordered distribution with all cards in the same suit and being in order from the ten to the ace. The probability of getting this is 1 in 650,000. Here the most random state is 325,000, more probable than the most ordered state. By the way, for you poker players, the probability of getting four aces is about 1 in 54,000, which is twelve times more probable than getting a royal flush. That is why a royal flush beats four aces.

Hopefully you see that the second law is not so surprising at all. It is just a statement that nature tends to the most probable or most likely state possible. While this sounds pretty obvious, it has extremely important consequences. A few will be discussed below.

- We should first consider the topic that started this section, the energy crisis. Let us use the example above about the lump of coal. Before being burned, the coal is in an organized, compact state of low entropy. After being burned, the coal is in the form of dust and distributed over a large area. It is in a much more randomized state, which is of higher entropy than initially. Energy has still been conserved, but it is now in a much more randomized form. To get that coal into a state that we would be able to use to extract the energy from it again would cost more energy than we originally obtained from it. It is usable energy we need for our societal needs. As we use it, we turn it into unusable forms. This is why we worry about an energy crisis. We are depleting the usable forms of energy.

- Let us consider another energy crisis, but this time on a universal scale. We can see what is happening on this scale by considering our sun, a very ordinary, typical star. The sun, of course, is our real source of energy. We receive this energy basically in the form of light energy. We only intercept a very small fraction of the energy the sun radiates. Most of it just goes out into empty space. Eventually the sun will use up all its nuclear fuel and burn out. What will have happened to all the sun's energy? It will have been dispersed throughout space, as the second law dictates. The energy that was once very intense and localized in the hot sun will be dispersed over the entire volume of the universe in a very cool state. For the vast majority of stars, this is their fate. So, in some very distant time (do not worry, it is like a few hundred billion years) the universe will be some cold and very dreary place. This process has sometimes been called the heat death of the universe. Not to cause you any undue anxiety, but it has been recently discovered that the expansion of the universe is actually accelerating, which means that the heat death will occur earlier than originally anticipated. Maybe its now tens of billions of years instead of a few hundred billion years.

- You have probably heard the statement "Heat always flows from hot to cold." Why does this occur, and what does it have to do with the second law? In order to best answer these questions, consider two volumes of gas, initially separated with one at a high temperature and the other at some lower temperature. We will look at the gasses on the molecular level. Temperature is a measure of the average kinetic energy of the molecules in any substance. Thus, the higher-temperature gas molecules have a higher kinetic energy, and hence a higher average speed. When the two gasses are connected, the hotter molecules having a higher speed will travel a farther distance than the cooler molecules. Thus, the heat energy flows from hot to cold. Also, by collisions, the higher-energy molecules lose energy to the lower-energy molecules. Thus, the hot molecules cool off while the cool molecules heat up. This process of energy transfer stops when all the molecules finally have the same average kinetic energy. In other words, the gas reaches a final, common temperature. In this case, the gas is in a more random state than when the hot gas molecules were all together in one region and the cool molecules were together in a separate region. The entropy has increased as the second law demands.

- Finally, let us imagine we can take a video of all of these processes described above: the lump of coal being burned and the coal dust being dispersed over some large area; the sunlight radiating from the sun, which eventually cools and dies out; and the hot gas molecules tending to travel to the cool ones, mixing and finally coming to a common temperature. Now consider running the video backward. What would we see? In all cases the scenes would be highly improbable: coal dust over the countryside somehow coming together in one location, forming a lump of coal; light from the far reaches of the universe heading toward one of a billion burned-out stars in one of billions of galaxies, hitting that one star and heating it to nuclear ignition; a gas at a uniform temperature separating into two gasses at different locations and at different temperatures. In all these cases we would immediately recognize that the video was indeed running backwards. In other words, we could tell when time was running forward or backward. Hopefully, this should sound fimilar. In Chapter 7, we discussed time reversal symmetry and the fact that it is violated on the macroscopic scale. We now see why. Since nature wants to go toward the most probable state, the state of highest entropy, this gives us an "arrow of time." Time "flows" in the direction of the increase in entropy.

But time reversal symmetry does still hold on the microscopic scale. Let us use our gas molecules to understand this. If we look at the gas molecules on any relatively short timescale, we see molecules colliding off each other. Running the video backward, we still see molecules colliding. But if we look at a good fraction of the molecules on a long timescale (the macroscopic scale), then we see the faster molecules separating from the slower ones, and all heading for a definite location. Definitely not natural.

8.4   END-OF-CHAPTER GUIDE TO KEY IDEAS

- How are symmetry and conservation related?
- What is the full statement of the conservation of momentum? What are three important aspects of this statement?
- What is meant by the system?
- What aspects of momentum conservation make it so powerful?
- How is momentum conservation expressed mathematically?
- What is the definition of energy?
- What important aspect of energy makes it so useful in society?
- What is the full statement of the conservation of energy? How is it expressed mathematically?
- What are the different forms of energy?
- Can you write the conservation of energy equation for the case of a ball while in the air or on a roller coaster?
- If energy is conserved, why do we seem to always be concerned about an energy crisis?
- What is the second law of thermodynamics? Can you state it in three different ways in terms of the three variables: entropy, randomness, and probability?

QUESTIONS/PROBLEMS

Note: In the problems below, where needed, use the 9.8 m/s² value for the acceleration due to gravity, g.

1. Why are conservation laws so important?
2. How are symmetry and conservation related?
3. State the conservation of momentum. Define all relevant terms.
4. In the conservation of momentum, why can't internal forces affect the
   momentum of the system?
5. Is momentum always conserved? Justify your answer.
6. Is momentum a scalar or vector quantity?
7. Write the conservation of momentum mathematically in two different ways.
8. Which of Newton's laws of motion are relevant in the conservation of
   momentum? Explain.
9. When a ball is thrown into the air, is momentum conserved? Justify
   your answer.
10. When a ball is first thrown into the air, what is the force that causes the
    earth to move downward?
11. When a ball is released from rest, what is the force that causes the earth
    to move?
12. A 12 kg bomb at rest explodes into two parts. An 8 kg piece flies off with a
    speed of 12 m/s to the left. Find the speed and direction of the other piece.
13. A 12 kg bomb is moving to the right with a speed of 10 m/s. It then explodes
    into two pieces, one of which has a mass of 8 kg and is moving to the left
    with a speed of 12 m/s. Find the speed and direction of the other piece.
14. A 12 kg bomb is moving to the right with a speed of 10 m/s. It then explodes
    into two pieces, one of which has a mass of 8 kg and is moving to the right
    with a speed of 12 m/s. Find the speed and direction of the other piece.
15. A 20 g piece of putty, moving with a speed of 1,000 cm/s, hits and sticks
    to a 500 g block at rest. After the collision, what is the speed of the putty-
    block system?
16. In any of the problems 12 to 14, is energy conserved? Is kinetic energy
    conserved? In each case, justify your answer.
17. Besides the fact that energy and momentum are different quantities, list
    three other ways in which they are different.
18. Write the conservation of energy in two different mathematical forms.
19. List and define the different forms of energy.
20. What is a conservative force? List as many conservative forces as you can
    think of.
21. Some old pendulum clocks required you to wind a spring in order for them
    to work. What is the purpose of the spring?
22. A super-ball is a special ball that is manufactured so it will bounce higher
    than most ordinary balls. Is it possible for a super-ball that is released (at
    rest) from your hand to bounce to height greater than from where it was
    released? Justify your answer.

23. A ½ kg ball is released at rest from your hand at a height of 10 m above the ground. What is the speed of the ball just before it hits the ground?
24. A ½ kg ball is thrown downward from a height of 10 m with an initial speed of 5 m/s. What is the speed of the ball just before it hits the ground. Do the same problem if the ball was initially thrown upward instead of downward.
25. An object is projected from the ground with a speed of 10 m/s. What maximum height does it reach?
26. An object is projected from the ground with a speed of 10 m/s. What speed does the object have while moving upwards at a height of 2 m? What speed does the object have when it is moving downward at the height of 2 m?
27. Choose which statements are true or false. For those that you choose as being false, explain why and correct them to make them true.
    a. If momentum is conserved, so is kinetic energy.
    b. A ball is thrown into the air. Its momentum is the same at the same height on the way up as on the way down.
    c. A ball is thrown into the air. Its kinetic energy is the same at the same height on the way up as on the way down.
    d. A ball is thrown into the air. Its momentum is conserved while it is in the air.
    e. In any explosion, momentum is always conserved.
    f. In a collision between two objects, momentum must be conserved.
28. If energy is always conserved, why are we concerned about an energy crisis?
29. State the second law of thermodynamics. Define all relevant terms.
30. Explain why the entropy of a system cannot decrease.
31. What does the second law of thermodynamics have to with time?

9 The History of the Atom

We have already mentioned the atom several times and have seen how it is key in understanding many things in nature. It is certainly one of the great unifying constructs in physics. The idea that all matter is made up of small, solid objects goes back some 2,500 years to the Greek philosopher Democritus. While we still certainly believe in the atom, it is quite remarkable that the picture (model) we have of the atom has changed only in the last 100 years. In this chapter we will concentrate on how and why that changed view has come about. It is a very nice example of the methodology in physics.

9.1 THE GREEK MODEL

This is the model of Democritus. Here the atom is a tiny, solid, indivisible, and indestructible object. In fact, the word *atom* comes from the Greek meaning "indivisible." This model was believed to be correct until the end of the nineteenth century.

What caused us to change our view of this picture? As in any other idea in science, we must abandon our old ideas when evidence is presented that is not consistent with those ideas. In the case of the atom, there were two different pieces of evidence.

The first was the discovery that atoms emitted light. Remember, in Chapter 6, we found out that light is produced when charged particles are accelerated. If the atom emitted light, then it had to be made up of accelerated charged particles. In fact, since the atom is electrically neutral, it had to be made of two oppositely charged objects. If that is the case, it cannot be solid and indivisible.

Also, in 1897, J. J. Thomson discovered the electron, which had a mass about 2,000 times less than the lightest atom. If the atom was supposed to be the fundamental building block of all matter, how could there be something smaller than it? Obviously the Greek model could not be correct. A new model was needed.

9.2 THOMSON'S "PLUM PUDDING" MODEL

How do we go about creating a new model? We start by using the facts that caused us to abandon the old model. For one, the new atomic model must have both positive and negative charges. At least one of the charges must be able to be accelerated to produce the light emitted from the atom. Also, it was observed that the atom did not normally radiate light, but only when it was excited by some outside force.

This leads to two possible models: a planetary model, where the light electrons circle a positive nucleus, and a plum pudding model, where the negative electrons are embedded in a positive pudding or jello-like material.

For the physicists of the time, the planetary model had to be rejected. The electrons, moving in circular motion, would be continuously accelerated. This means they would be continuously radiating electromagnetic energy, which was not consistent with the last observation discussed abve. Also, a quick calculation shows that if the electron is continually radiating energy, it would spiral into the nucleus in about 10^{-6} s. In other words, the atom would destroy itself in that amount of time. Since we know atoms are stable (after all, the earth has been around for about 5 billion years), it appeared the planetary model could not be correct.

It was J. J. Thomson who suggested the plum pudding model. In his model the atom is stable since the electrons are normally at rest embedded in the positive pudding-like medium. The atom can radiate light if the electrons are caused to accelerate by some outside force. The electrons radiate as long as this outside force is applied, and then quickly stop since the viscous type pudding would damp out the motion quite rapidly. This agreed very nicely with observation, which was that the atom would radiate only when excited by some outside disturbance and stop emitting light once the disturbance was removed. So this model sounded pretty good, except for a bothersome detail.

The frequency or wavelengths of the atomic spectra could easily be calculated in this model. In fact, the prediction is just like the spectra of musical notes from, let us say, a violin. Namely, there should be a lowest or fundamental frequency, and then harmonics, which are just integral multiples of the fundamental. Unfortunately, this did not agree with the observed spectra, which were very different from the above prediction.

The plum pudding model seemed so nice that physicists at the time did not give up on it, but rather tried to "patch it up." They reasoned that the pudding somehow acted in a more complicated way on the electrons than envisaged by the original, simple model. By the way, this is very typical of what physicists do when they have a theory that they like but does not quite agree with observation; they try to find some way of patching it up. Unfortunately, the observed spectra were so different than the prediction, no one was able to find a model that worked. But the real blow to the plum pudding model came in 1911 by a now famous experiment by Ernest Rutherford.

9.3   THE RUTHERFORD EXPERIMENT

The idea of the Rutherford experiment is very simple. Imagine you have some pudding (or jello, if you prefer) with plums (no pits) embedded throughout it. You now shoot a bullet through the plum pudding and note what happens to the bullet after it passes through. What you would expect is that the bullet would pass through with very little, if any, deflection. The pudding or the plums should have essentially no effect on the heavy, fast-moving bullet.

Instead of a 22-caliber-type bullet, Rutherford used atomic-type bullets known as alpha particles. These were known to be positively charged and have a mass four times that of the hydrogen atom (we will have a furter discussion of alpha particles in the next chapter). The experiment was to shoot his alpha particle bullets through a very thin foil of gold. The alpha particles could be detected since, when they hit a flourescent screen, they would cause the material in the screen to emit a flash of light. Rutherford surrounded the foil with the flourescent screen so he could detect alpha particles coming off in any direction after passing through the gold. His expectation, if the plum pudding model was correct, was that the alpha particles would pass through the gold foil undeflected, as depicted in Figure 9.1a. To his surprise, some of the alpha particles came off with large deflection angles, as shown in Figure 9.1b. Some even were deflected backwards, indicating they had to bounce back off something heavy in the gold atoms.

This was totally inconsistent with the plum pudding model, indicating it could not be the correct model for the atom. Before going further with our story, it should be noted that the information about the structure of the atom has been obtained without ever being able to see the atom. We have observed either what the atom emits (in the form of light) or what affect the atom has on something else. In other words, the observations have been indirect in contrast to direct sensory observations. This is, in fact, the nature of most scientific observations. Very rarely do we get direct sensory information. We will see this again and again in this book.

9.4   THE PLANETARY MODEL

The fact that some of Rutherford's bullets bounced back from the gold atoms meant that the alpha particles had to be hitting something more massive than they were, and something with a positive charge that would repel the positive charge of the alpha particles. A marble striking a bowling ball must bounce back off the bowling ball. On the other hand, a bowling ball striking a marble will just continue in the same direction. According to the laws of conservation of momentum and energy, it is impossible for a more massive object to bounce back off a less massive object.

If the atom is made of massive, concentrated, positive objects, the plum pudding model could not be correct. This then leaves only one other choice, the planetary model. In this model, remember, the light electrons circle a positively charged massive nucleus. This picture is consistent with the observations in the Rutherford experiment. Unfortunately, as discussed above, this model contradicted the laws of physics that were believed to be true at that time.

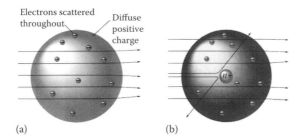

FIGURE 9.1   (See color insert following page 102.) (a) If the plum pudding model were correct, alpha particles would pass through the atom with little or no deflection. (b) What Rutherford observed was that some alpha particles were deflected through large angles, indicating the presence of a massive nucleus. (From Ted Ankara; library.tedankara.k12.tr/chemistry/vol3/vol3.html)

9.5 WHAT DO WE DO NOW?

The results of the Rutherford experiment created a serious dilema for physicists at that time. The solution to understanding the atom has to do with the ideas of quantum mechanics that began in 1900 and were then reinforced by Einstein in 1905. We will have to wait to a later chapter to get into some of the detailed ideas associated with the quantum theory. But it is instructive here to relate part of the next important step in the story.

In 1913, a young Danish physicist, Neils Bohr, proposed some revolutionary ideas. Some were based on the newly emerging quantum ideas, but they were not derivable from any previously known theory. These type of ideas, or assumptions, are called *ad hoc* assumptions. The words *ad hoc* are Latin meaning "made up at the time." Some of you may have served on an *ad hoc* committee, which is one that is formed at the time to deal with a specific situation. In science, *ad hoc* assumptions are hypothesized if observations contradict previously believed theories, so brand new ideas have to be formulated. This is usually when the greatest revolution in scientific thinking occurs.

Bohr used his new ideas to predict the spectrum of light emitted by hydrogen, the simplest of all atoms. Remember, the plum pudding model had a predicted spectrum that did not agree at all with the observed one. But Bohr's prediction agreed exactly with the observed hydrogen spectrum. Obviously, he was on the right path, which suggested that these new quantum ideas, which seemed very strange at the time, had to be taken seriously.

While we have to stop our story of the atom here, I hope you see how the interplay of theory and experiment led to our present understanding of the atom. It is the essence of scientific methodology.

9.6 THE ATOM TODAY

While the description of the atom we have presented may seem rather strange, the fact is that we do understand the atom very well. Without this understanding, we could not understand modern chemistry or modern molecular biology. There would not be the great strides we have made in our knowledge of DNA and genetics. Although we cannot go into the modern theory of the atom here, there are some detailed facts that will be instructive to review now.

Our picture is a modification of the planetary model. There is still a positively charged nucleus in the center, but the electrons do not orbit the nucleus like a planet moving around the sun. Rather, the electrons form a cloud of charge. This is a fundamental change of our picture of the atom. It takes a knowledge of quantum mechanics to better understand this. For hydrogen, the atom contains one proton with one electron. The average radius of the electron cloud, and hence the size of the hydrogen atom, is approximately 10^{-8} cm. For other atoms, there are more electrons, and their average radii are correspondingly larger. The force that holds the atom together, as we have said before, is not some special atomic force, but just the electrical attraction between the positive nucleus and the negative electrons.

The electrons can only be in very well-defined, discrete energy levels with corresponding discrete average radii. Normally, the electron is in a lowest-energy state and does not emit any radiation when in this state. This, as was said earlier, is not consistent with classical physics but is a consequence of the quantum nature of the atom. The atom can only emit radiation when the electron makes a transition from a higher energy level to a lower one. This can only happen if the electron is first raised to one of these higher-energy states by some outside excitation, for instance, by passing an electrical current through the atoms of a particular material or heating it to high temperatures.

You may be wondering why the electron has to make a transition from a higher- to a lower-energy state. Why doesn't it just stay in the higher-energy state? Well, we do not need quantum theory to explain this. It can be called an energy prime directive:

All systems want to go the lowest potential energy possible.

Remember, that is because there is a force trying to reduce the potential energy. For instance, the reason something falls toward the ground is because the gravitational pull of the earth is attracting it there, thus reducing its potential energy. When the atom is in an excited state, it has a higher potential energy, and hence will want to lower it due to the electrical attraction between the positive nucleus and the negative electron. The nonclassical part is when the atom is in its lowest energy state, which is not when the electron is at the nucleus. It is the quantum nature of nature that prevents it from going to a lower energy.

Because the energy levels are discrete, the differences between levels are also discrete. Thus, the energy of the emitted radiation must be discrete. Since the frequency of the radiation is proportional to the energy, the frequencies from a particular atom, the atomic spectra, are unique and discrete. It was the agreement between the predicition of these discrete frequencies of the Bohr model and the observed spectra that caused such a sensation at the time.

Each atom has its own set of discrete spectra. Thus, atomic spectra act as fingerprints to uniquely identify what atoms are present in any substance. Some of you, at some time, may have wondered how we know what the sun or even more distant stars are made of. Well, hopefully you realize that the statement above gives the answer. We, of course, see the light coming to us from a star. By the use of something known as a diffraction grating (which acts like a prism in that it causes different frequencies to be deflected to different angles), we can observe the different atomic spectra, and hence identify the different elements in the star.

Interestingly, there is one element that was detected on the sun in 1868 before it was ever observed here on earth. In fact, that element is named after the sun. The Greek word for sun is *helios*, and the element is helium.

A question: If we had not observed helium on earth, and thus could not know what its atomic spectrum was, how could we identify it on the sun? Try to come up with the solution before reading on.

The answer: By the process of elimination. When observing the full spectrum from the sun, the spectra from all the known elements could be eliminated. When this was done, there were still a set of discrete frequencies left over. Thus, an unknown element had to exist on the sun that had not yet been observed on earth. Helium was finally detected on earth in 1895, when the spectrum from an unknown gas was shown to be the same as that observed from the sun.

9.7   THE ELECTRON VOLT: A USEFUL ENERGY UNIT

Above, we have talked about the energy levels in the atom, but did not mention how big these energies are. For instance, the binding energy of the electron in the hydrogen atom (the energy needed to free the electron from the nucleus) is 2.18×10^{-18} Joules. This certainly looks like a very small number, but on the atomic scale, we really have no physical feeling as to how large, or small, this really is.

When we refer to atomic and subatomic energies, it is much more useful and convenient to use a new energy scale known as the electron volt. It is defined as the energy an electron (or proton that has the same magnitude of electric charge) would gain in going through a potential difference of 1 V. Remember, batteries are graded in volts. For instance, flashlight batteries are 1.5 V and car batteries are 12 V. When an electron is accelerated from the negative to the positive terminal of a 1 V battery, it gains an energy of 1.6×10^{-19} Joules. We call this amount of energy 1 electron volt, or 1 eV.

$$1 \text{ eV} = 1.6 \times 10^{-19} \text{ Joules}$$

In these units, the binding energy of the electron in the hydrogen atom is 13.6 eV; i.e., it is the energy gained in being accelerated by a 13.6 V battery.

To give some more feeling as to the electron volt unit of energy, you might find the following table useful.

Energy of:	Energy in eV Units
Flashlight battery	1.5
Visible light	1.5–3.0
X-rays	$\approx 50,000$
Gamma rays	$\approx 1,000,000$ (1 MeV)
90 mph baseball	3×10^{15} MeV

When we get to larger energies in eV units, it is convenient to introduce a new unit, the MeV (10^6 eV), where the M stands for million. There are other units, such as GeV, which is equivalent to 10^9 (1 billion, or giga) eV.

It is also very convenient to express the masses of elementary particles such as the electron and proton in eV units. The reason we can do this is because of Einstein's famous equation $E = mc^2$, which relates mass and energy. In this equation, E is energy, m is mass, and c stands for the speed of light. Below is another table giving the masses of the electron, proton, and neutron in both kg and MeV units.

Particle	Mass in kg	Mass in MeV
Electron	9.11×10^{-31}	0.51
Proton	1.673×10^{-27}	938
Neutron	1.675×10^{-27}	939

Several things should be noted about the masses in the above table. First, the proton and neutron masses are almost 2,000 times that of the electron mass. Thus, just about the entire mass of the atom is contained in the nucleus. Second, the neutron mass is slightly larger than the proton mass. The difference is only about 1 part in 1,000. These facts will be important in our discussion of the atomic nucleus in the next chapter.

9.8 END-OF-CHAPTER GUIDE TO KEY IDEAS

- What was the Greek model of the atom? Why was it finally rejected?
- What two models were proposed to replace the Greek model? Can you describe each one?
- Why was one of them initially selected over the other?
- What was one difficulty the plum pudding model had from the start?
- Why was the plum pudding model finally rejected?
- Can you describe the planetary model, especially as we view it today?
- What was Bohr's contribution to this picture?
- What is the energy prime directive?
- What is an electron volt? Why is it a useful unit of energy?
- What is the binding energy of the electron in Joules and electron volts?
- Why can we quote masses in electron volts?

QUESTIONS/PROBLEMS

1. List and describe the three models of the atom discussed in this chapter.
2. Describe the Greek model of the atom.
3. Why was the Greek model of the atom rejected?
4. What two models of the atom were proposed after the Greek model was rejected? What did these models have in common? What were the differences?
5. What was the main reason that the Thomson model was initially accepted over the planetary model?
6. In the planetary model, what was the process by which the atom emitted light?
7. In the Thomson model, what was the process by which the atom emitted light?
8. In the Thomson model, what was the predicted nature of the atomic spectra? How did this compare with the observed spectra?
9. What was the purpose of the Rutherford experiment?
10. Describe the idea of the Rutherford experiment. What was expected?
11. What was observed in the Rutherford experiment? Why was this not expected for the Thomson model?
12. Why was the result of the Rutherford experiment consistent with the planetary model?

13. What is an *ad hoc* theory?
14. Describe the picture we have of the atom today.
15. What is the energy prime directive?
16. Why does an electron, in an excited state, fall to a lower energy state?
17. Describe the nature of the observed atomic spectra.
18. Why are atomic spectra discrete?
19. How do we know what distant stars are made of?
20. Define the *electron volt*.
21. Why is the electron volt a convenient unit to use for atomic energies?
22. How much energy, in Joules, is required to free the electron from the proton in the hydrogen atom?
23. Show that a baseball with a speed of 90 mph has an energy of about 10^{15} MeV. Guesstimate a reasonable mass for a baseball. You will have to use MKS units to get the correct numerical answer.

10 The Nucleus

As we have seen, the atom is the key to understanding all the forces, except for gravity, that affect us in our everyday life. For these atomic interactions, the nucleus plays a dual role. It holds the atom together by the electric attraction between the positive constituents in the nucleus and the negatively charged electrons. It also attracts the electrons in other atoms to cause these atoms to stay together to form the matter that we come in contact with, including ourselves. We have seen earlier that, in fact, there is a delicate attractive-repulsive balance causing atoms to attract each other in some cases, and in others causing repulsion, as in the case, for example, when a knife cuts through butter. In addition, except for the lonely proton in hydrogen, the nucleus has a rich and complex life of its own.

10.1 NUCLEAR PROPERTIES

The nucleus is made up of protons and neutrons. Protons have a positive electrical charge, while neutrons, as the name implies, are neutral. As we noted at the end of the last chapter, the proton and neutron masses are almost identical, being approximately 2,000 times the mass of the electron but differing by only about 0.1%. A typical nuclear size is about 10^{-14} m, about 1/10,000 (10^{-4}) the size of the atom.

> A question? These facts should suggest a question in your mind. As usual, try to think of it before reading further.

The question: What holds the nucleus together?

A first guess might be the gravitational force. After all, it is always attractive, and over such minute distances perhaps it is strong enough to bind the nuclear particles together. But unfortunately, gravity cannot do it. In Chapter 5, we showed that the ratio of the electric force to the gravitational force is 10^{42}. In other words, the protons are repelling each other with an electric force tremendously larger than the attractive gravitational force. Thus, there must exist an attractive force greater than the electrical repulsive force. It is the strong nuclear force. It acts over very short distances, about 10^{-15} m, which is the separation between protons and neutrons in a nucleus. The strength of the force is the same between two protons, two neutrons, or between a neutron and proton. Also, while it is indeed stronger than the eltrical force, it is only about ten times stronger, not the huge factor of 10^{42}, as in the comparison between electricity and gravity. This has important consequences, as we shall discuss below.

The properties of any nucleus are uniquely determined by the number of protons and neutrons. We will designate a nucleus by the following three symbols:

Z (atomic number) = Number of protons

N (neutron number) = Number of neutrons

A (atomic weight) = Z + N

A, of course, is not a weight, but a number. It is called the atomic weight since the mass of the atom is essentially determined by the number of protons and neutrons given the fact that they both have a mass approximately 2,000 times that of the electron. If we are not interested in distinguishing between a proton and neutron, we use the word *nucleon*. Then, A gives the number of nucleons in the nucleus.

In terms of the symbols above, a nucleus is designated as ^AX_Z, where X is the symbol for that particular element. For instance, $^1H_1$ designates the proton that is the nucleus of hydrogen; $^4He_2$ represents the helium nucleus, with 2 protons and 2 neutrons; and $^{238}U_{92}$ is the symbol for uranium, with 92 protons and 146 neutrons. This is a redundant notation since the number of protons uniquely determines the element, but it is the standard notation and we will stick with it.

It is the atomic number, Z, that determines just about all the chemical properties of an element. It is possible, though, to have nuclei with the same atomic number but different neutron numbers. Atoms (or nuclei) with the same Z but different N are called isotopes. Atoms of different isotopes do behave the same chemically, but the nuclei do have different properties, especially when we consider nuclear radioactivity, which we will discuss below.

10.2 WHY NEUTRONS?

As was noted above, the protons in the nucleus both hold the atom together and attract the electrons from nearby atoms to cause atoms to stay together. So, why does nature need neutrons? The answer has to do with the strength of the nuclear attraction compared to the strength of the electric repulsion between protons and the range of the nuclear force. While the nuclear attraction between any two protons is about ten times the repulsion, we have to consider what happens with more than two protons.

Let us consider a nucleus with some number of protons. Because of the very short range of the nuclear force, a given proton is only attracted by other protons in its immediate vicinity. So, all of the other protons in the nucleus do not partake in the attraction. This is known as saturation of the nuclear force. On the other hand, due to the very long range of the electric force, all other protons in the nucleus contribute to the repulsion. Thus, it would not take that many protons before the repulsion would win out and only nuclei with relatively small atomic numbers could exist.

Neutrons save the day in two ways. First, they provide attraction due to the nuclear force between them and the protons without adding any repulsion. Second, they cause the protons to be farther apart, on average, which does lessen the repulsion since the electric force does fall off with separation. We see evidence for these effects in Figure 10.1, which shows the proton and neutron number for all known isotopes. For the lighter elements, the number of protons and neutrons tends to be equal, while for the heavier elements, there are more neutrons than protons. For example, the most

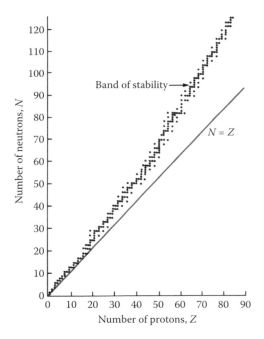

FIGURE 10.1 Neutron number versus atomic number for the known stable nuclei. For those nuclei above the N = Z line, there are more neutrons than protons. (From Sean Putnam; lhs. lps.org/staff/sputnam/chem_notes)

common isotope of carbon is $^{12}C_6$, with six protons and six neutrons. On the other hand, the most common isotope of uranium is $^{238}U_{92}$, with 92 protons and 146 neutrons. In fact, there are no stable isotopes for any element with a Z larger than 82, which corresponds to the element lead. Without neutrons we could not have any of the heavier elements, which are so necessary for our existence.

10.3 NUCLEAR DECAYS

You have probably heard something about radioactivity. It has both good and bad aspects. For instance, in medicine, radioactive isotopes are used to treat certain cancers. On the other hand, the problem with nuclear waste has to do with the danger of high levels of radioactivity in the waste material. In all cases, radioactivity comes about because certain isotopes undergo nuclear decay where an energetic particle is emitted and a new nucleus remains that, in some cases, has a different atomic number compared to the original nucleus. In other words, the new nucleus is that of a different element.

It should be noted that not all nuclei undergo decay. Some are perfectly stable. Why should some nuclei decay and others not? Basically because of the energy prime directive. If, by decaying, the new nucleus is in a more stable, lower-energy state, then it will do so. It is the nuclear analogy of the radiation emitted in atomic spectra. Remember, the atomic radiation is always electromagnetic, but because the nucleus is more complex, there are three different types of radiation that can be emitted. They are known as alpha (the Greek letter α), beta (β), and gamma (γ).

10.3.1 ALPHA DECAY

Alpha decay occurs when an α-particle is emitted from a heavy nucleus. An α-particle is a $^4\text{He}_2$ nucleus. After emission, the new, remaining nucleus has two less protons and four less nucleons (two protons + two neutrons). We know this because of conservation of charge (the α-particle has carried away two positive charges) and conservation of matter (it has carried away four nucleons). The new nucleus is more stable due to the fact there are fewer protons, and hence less electric repulsion. As we have seen above, it is the heavier nuclei that are the most unstable.

We can write a nuclear decay equation analogous to a chemical reaction equation:

$$^A X_Z \rightarrow \alpha + {}^{(A-4)}Y_{(Z-2)}$$

where X is the symbol for the element with atomic number Z and Y is the symbol for the element with atomic number $(Z - 2)$. For example, for α-decay of uranium, we write:

$$^{238}U_{92} \rightarrow \alpha + {}^{234}\text{Th}_{90}$$

We know the new nucleus is thorium (Th) since that is the element with atomic number 90. Thorium is also unstable and thus radioactive. In fact, all the subsequent daughter nuclei decay by α-emission until lead is reached, which is stable.

Alpha radioactivity is particularly dangerous since the α-particle has two units of charge and is much more massive than other forms of radioactive particles. Thus, they can easily strip (ionize) electrons from atoms with which they come in contact. For instance, in our body, they can interact with water atoms (H_2O) to form the OH ion (called a free radical). These are highly interactive and can cause damage to cells and especially the DNA. It is believed that this type of damage to the DNA can then cause the formation of cancer cells.

10.3.2 BETA DECAY

A β-particle is an electron emitted from a nucleus.

Does this statement evoke a question?

Hopefully you have asked yourself: If there are only protons and neutrons in the nucleus, how did an electron get there? The answer has to do with one of the interesting (weird?) properties of subnuclear particles. Not only do nucleii decay, but subnuclear particles like neutrons can also decay. The decay is like an explosion where the initial (parent) particle disappears and the final (daughter) particles are the remnants of the explosion. They were not in, or part, of the original particle. They are created as a result of the decay process.

In the case of β-decay, we can picture a model that is consistent with a neutron in the nucleus decaying into an electron and proton. If this is correct, we can write

$$n \rightarrow p+e$$

Note that this is consistent with conservation of charge since the neutron has no electric charge and the charge of the proton (+1) and the charge of the electron (−1) add up to zero. It is also consistent with the conservation of energy. If we look back at the table of mass-energies at the end of the last chapter, we see the mass of the neutron is 939 MeV, while the proton and electron masses are 938 and 0.51 MeV, respectively. It might at first appear that 0.49 MeV of energy has been lost, but remember the electron is emitted from the nucleus, and so the extra energy can be understood as coming in the form of kinetic energy of the emitted electron. You may wonder why the electron and proton do not share the energy equally. After all, in a two-body explosion at rest, both particles must come off with equal and opposite momenta. The answer is in the fact that while momentum and kinetic energy both have to do with motion, they are not the same thing, as we show below:

$$KE = \tfrac{1}{2}mv^2 = \tfrac{1}{2}mv^2(m/m) = \tfrac{1}{2}(mv)^2/m = p^2/(2m) \tag{10.1}$$

This expression tells us that for two objects with the same momentum but different masses, the object with the greater mass will have the lower kinetic energy. Since the proton has a mass about 2,000 times more than the electron, it takes away a very small fraction of the energy even though it has the same momentum. Since the proton has so little energy after the decay, it remains in the nucleus.

We can now write the nuclear decay equation:

$$^{A}X_Z \rightarrow \beta + {}^{A}Y_{(Z+1)}$$

Do you have some questions here?

There could possibly be two questions about the above equation. One has to do with the atomic weight. Some of you may already see why it has to be the same for both the parent and daughter nucleus. If not, see if you can first understand it on your own. Let us go through the explanation to make sure. Since the neutron decayed into a proton, the number of nucleons does not change, and hence A does not change.

The other question you may have should deal with the fact that the final nucleus has a larger electric charge. Why should that be? After all, we have implied before that more charge would make the nucleus less stable, not more. Yet, a decay occurs only if the final state is more stable than the intial one. The answer has to do with the subtleties of nuclear physics. The nucleus is quite complex. For example, in calculating the energy of a given nucleus there are five different terms that are relevant;

some increase the energy and others decrease it. So, there can be a delicate balance. In some cases, the energy will be lower, with more instead of fewer protons.

This is a good example of the importance of questioning. If physicists did not stop to question why a higher charge meant a more stable state, we might not have understood an important aspect of the physics of the nucleus.

10.3.3 GAMMA DECAY

A gamma particle is electromagnetic radiation emitted from a nucleus. In this case, a nucleus loses energy but does not change its atomic number or atomic weight. In some cases, the nucleus has become excited because it has just undergone alpha or beta decay. So to speak, it is happy with its "personality" but can get to a more stable state by just emitting energy. The nuclear decay equation can be written as

$$^AX^*_Z \rightarrow \gamma + {}^AX_Z$$

where X^* designates an excited nucleus.

Gamma decay is most similar to what happens when atoms emit atomic spectra in that, in both cases, electromagnetic radiation occurs. But in the case of nuclear gamma emission, the energy is 10,000 to 100,000 times greater. In fact, the energies involved in just about all nuclear processes are this much greater than the energies involved in atomic, thus chemical processes. This is why the fuel needed in a nuclear power plant is just inserted at the start-up of the plant and does not have to be replenished for about 30 years, while in a coal-burning plant, a trainload of coal is usually delivered every few days.

10.4   HALF-LIFE AND CARBON DATING

All nuclear decays occur over some finite time interval. In other words, all nuclei of a given species do not just decay all at once. The time evolution of these decays is not unique to just nuclear processes but is, in fact, quite common throughout nature. As an example from biology, the same time dependence can be observed in osmosis, where the concentration of some liquid changes as it permeates through the wall of some cell membrane. Decays of this nature are described mathematically as exponential decays. A very useful and unique time associated with these forms of decay is known as the half-life.

Half-life is the time it takes for one-half of a substance to decay. At the end of one half-life, there is one-half of the original substance left. If a second half-life goes by, there will be one-half of what you started with at the end of the first half-life. In other words, one-fourth $(1/4) = (1/2)^2$ of what was originally there. After three half-lives, there would be $(1/2)^3 = 1/8$ of the original amount. This is depicted graphically in Figure 10.2.

Nuclear half-lives have a very large range, from smaller than 10^{-6} s to over a billion (10^9) years. Each radioactive nucleus has its own well-defined half-life. If we know the half-life of a particular nucleus, we can use it as a clock. This is the basis for radioactive dating, which is the method of determining the age of a substance by measuring what is known as the decay rate in that substance. Carbon dating is probably the best known of the dating techniques, although it is not the only one used.

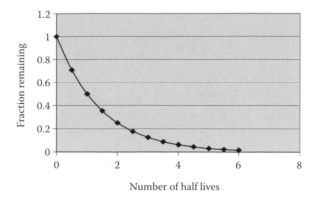

Number of half lives

FIGURE 10.2 Graph showing an exponential decay indicating the fraction of the original number left after different numbers of half-lives.

Carbon-12 (^{12}C) is by far the most common isotope of carbon. It is very stable and hence does not decay at all. On the other hand, ^{14}C is very rare, with only one nucleus in about 10^{12} being the ^{14}C isotope. But it does undergo radioactive decay with a half-life of 5,700 years. Let us remember what that means—if you start off with a certain amount of ^{14}C, after 5,700 years, half of what you started with would have decayed. After another 5,700 years, one-fourth of the original amount would be left, etc.

Now let us turn this around a bit. If the amount of ^{14}C in a sample can be measured today and it is known how much ^{14}C there was originally, then we know how old the sample is. For instance, if we know that there was originally 1 g of ^{14}C in a sample initially and we now measure ½ g, we know the substance is 5,700 years old. If we measure ¼ g, then we know the substance is $2 \times 5,700 = 11,400$ years old.

Do you have some questions here? You should have at least two.

First, if there is so little ^{14}C in a sample of carbon, how can it ever be measured? Second, even if the amount of ^{14}C can be measured now, how can it be known how much there was originally over thousands of years ago?

To answer the first question, we have to realize that we do not *directly* measure the amount of ^{14}C by, for instance, weighing the sample. The fraction of ^{14}C is just too small. So we must do it *indirectly*, as we do many measurements in science. In the case of radioactivity, the rate of decay (the number of decays in a specific time interval) is directly proportional to the number of decaying nuclei. In other words, if the number of nuclei is reduced by, let us say a factor of 2, then the rate of decay is also reduced by a factor of 2. Thus, we have to compare the rate of decay of ^{14}C now to the original rate of decay. We never have to know the actual number of ^{14}C nuclei. We just have to measure the decay rate. So, if we look, for example, at a living tree today and measure the present decay rate of ^{14}C found in 1 g of carbon from that tree, there will be 16 ^{14}C nuclei decaying in 1 min—the decay rate is 16/min/g. Note this also implies that if there was ½ g of carbon, the rate would be 8/min, and with 2 g, the rate would be 32/min, etc.

But we still have the problem of knowing the original decay rate thousands of years ago. In order to understand this, we have to understand how ^{14}C is produced. It is not a naturally occuring isotope on earth. It is produced by cosmic rays, which are particles coming in from outer space. In the case of ^{14}C, it is produced when cosmic ray neutrons hit and interact with the most common element in the air, nitrogen. The nuclear reaction equation can be written as

$$n + {}^{14}N_7 \rightarrow {}^{14}C_6 + p$$

In words, when a neutron interacts with a nitrogen nucleus in the air, it produces ^{14}C plus a proton. As far as we can tell, this is the only way ^{14}C is produced on earth. Chemically, ^{14}C is identical to ^{12}C. When anything takes in carbon, trees or animals, for instance, it will have a minute amount of the ^{14}C along with the naturally occuring ^{12}C.

We have measured the amount of ^{14}C that is now being produced. But how does that tell us how much of it was produced in the distant past? Well, if the number of cosmic ray neutrons was the same in the past as it is now, then the production rate of ^{14}C would have been the same then as it is now. Wow, that sounds like a pretty big assumption! It is. But we have good reasons to believe that it is a fairly accurate assumption. One reason has to do with the age of the universe. Cosmic rays that impinge on the earth come from the cosmos (i.e., not only from our own solar system but from distant galaxies—from the universe). The universe has been measured to be about 13.5 billion years old. So 1,000 or even 100,000 years is a minute fraction of the age of the universe. Over such a small time, we might expect only a small change, if any, in the number of cosmic rays hitting the earth. By the way, the earth is about 5 billion years old, so thousands of years is also a small fraction of the earth's age.

The above argument sounds reasonable, but does not constitute *proof* that carbon dating is accurate. Can we verify the method by some independent means? The answer is yes. In fact, there are two methods that are both independent of each other and carbon dating. One is dating by counting tree rings, and the other is by uranium dating. Neither of these has anything to do with cosmic rays, and both agree with each other for ages that both can be compared (for objects between 5,000 to 12,000 years old). For objects less than about 12,000 years old, carbon dating agrees with the other two. But for objects about 20,000 years old, carbon dating appears to give an age about 10% to 15% lower than uranium dating. So, it seems our assumption about the constancy of the cosmic ray rate is not quite correct, but is not terribly wrong. And in fact, since we know the tree ring and uranium data, we can even correct the ^{14}C times.

The discussion in the above paragraph is an important one. It shows that we must be aware of our assumptions and, if we can, try to find ways of verifying them. This is as important in everyday life as in science.

Before we go on to the next topic, let us see how carbon dating works by an example. If we have a wood carving, for example, that contains 1 g of carbon and has a decay rate of 4/min, we get the age by first calculating what fraction this is of the original decay rate. For 1 g, the original rate is 16/min, so the fraction is

¼ = (1/2)². A time of two half-lives, or 11,400 years, has transpired since the tree that the wood was in had died—hence the age of the carving is 11,400 years. This is still an accurate date according to our uranium/tree ring calibration. As a bit more complicated example, consider a carving with 2 g of carbon with the same decay rate of 4/min. Since 2 g has a rate of 4/min, 1 g would have a rate of 2/min. Now we can compare to the original rate of 16/min/g, which is 2/16 = 1/8 = (1/2)³. We conclude the age is three half-lives, or 17,100 years. But we also know that the carbon dating disagrees with uranium dating by about 10% in this period. So, the accurate age is 18,800 years.

10.5   THE FULL BETA DECAY STORY

Why go back to beta decay? Because we have not discussed the full story. There are important parts missing. More importantly, the full story is an excellent example of both the methodology of physics and the use of conservation laws.

10.5.1   THE PREDICTION

We begin by applying the conservation of energy and momentum to the picture we had earlier, namely, a neutron decaying into a proton and electron. Since this is a two-body decay, just like a bomb exploding into two parts, as we discussed in Chapter 8, the proton and electron both come off back-to-back with the same momentum, as shown below.

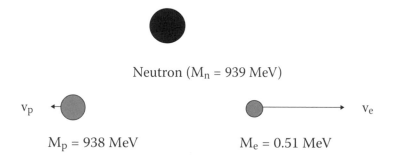

FIGURE 10.3   (See color insert following page 102.) A picture of beta decay where a neutron decays (explodes) into a proton and an electron.

Note the velocities of the proton and electron are also shown, indicating that the proton has a much smaller velocity than the electron even though they both have the same momentum. More importantly for our purposes, this also means that the kinetic energy of the electron is much greater than that of the proton, as we discussed in Section 10.3.2 on beta decay (remember we can write $KE = p^2/2m$). Using this, let us write a conservation of energy equation:

$$\text{Energy (after decay)} = \text{Energy (before decay)} \tag{10.2}$$

where energy can be in the form of mass-energy or kinetic energy. Writing the mass-energy for each particle along with the total kinetic energy released after the decay, we have:

$$938 + 0.51 + KE = 939 \text{ (energies in MeV units)} \qquad (10.3)$$

or

$$KE = 939 - 938 - 0.51 = 0.49 \text{ MeV} \qquad (10.4)$$

And since the proton has such a small fraction of the kinetic energy, the electron is emitted with a kinetic energy of 0.49 MeV.

This is an important result. To review, by applying the conservation of energy and momentum, we predict that all electrons emitted in beta decay should have a kinetic energy of 0.49 MeV. We represent our prediction graphically in Figure 10.4.

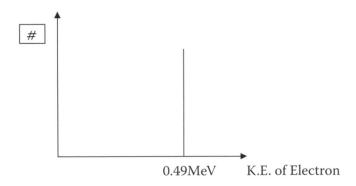

FIGURE 10.4   Graph of predicted energy distribution of emitted electrons in beta decay. The vertical line indicates that all electrons are predicted to be emitted with the same energy of 0.49 MeV.

Is this a correct prediction? Well, in science we know how to answer this: we do an experiment to verify our prediction.

10.5.2   THE EXPERIMENTAL RESULTS

In this case, conceptually at least, the experiment is quite simple. We just need some material containing nuclei that undergo beta decay and some equipment to measure the energy of each electron. The details of the apparatus are not important here. The results of such an experiment can be summarized graphically, as shown in Figure 10.5 in the smooth curve, which is superimposed on the graph of the prediction.

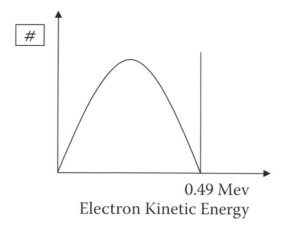

0.49 Mev

Electron Kinetic Energy

FIGURE 10.5   Graph of experimentally measured distribution of kinetic energies of electron emitted in beta decay. These results are superimposed on Figure 10.4, which shows the predicted distribution of energies.

> Before going further, try to read the experimental curve to see what it is telling you

10.5.3   What Do We Do Now?

By looking at Figure 10.5, it is obvious the measurement does not agree with the prediction. What do we do now? Basically, we start to scratch our heads and think of possible explanations. You might think the first thing a physicist would conclude is that the prediction (theory) is wrong. But physicists, like most people, are fairly conservative and do not like to throw out well "loved" theories such as conservation of energy and momentum. Before going down that path, let us see what else could be wrong.

Well, if we want to keep the theory, then the experiment must be wrong. In fact, when doing any experiment, a scientist must be on eternal vigilance to ensure that errors of one sort or another do not corrupt the results. This is especially true if the experiment seems to disagree with accepted theory. So, the first thing a scientist will do is repeat the experiment, checking out as many things that he or she can think of. If the results are the same, the scientist will submit an article for publication detailing both the experimental technique and the results. This article serves two purposes. First, it says to other scientists that an unexpected result contradicting accepted theory has been obtained. Please do an independent experiment to check it out. Also, if my original results are indeed correct, I will be glad to accept the Nobel Prize. After all, scientists are human beings who like recognition, rewards, and accolades like anyone else.

Let us discuss one subtle way that the experiment could be wrong. If we look at Figure 10.5, we see that the measured energies are less than predicted. Remember that the nuclei where the beta decay is taking place are imbedded in some material. It is possible that in traversing the material (i.e., going past literally billions and billions of atoms) to finally be detected in the experimental equipment, the electrons have lost some energy. In other words, the true energy has not been detected, but one that has been reduced by interactions with the atoms of the material. This is a perfectly reasonable hypothesis, which can also be tested. Before reading further, see if you can think of a way of checking this out.

If the problem has to do with energy loss in going through material, then the amount of energy loss can be modified by changing the amount of material. If less material is used, then there will be less energy loss and the curve in Figure 10.5 will shift closer to the predicted value of 0.49 MeV. If more material is used, then the resultant curve will shift to even lower energies. This is a definitive test. When this test was performed, there was essentially no effect on the experimental results.

The experiment is correct. Hence, the "theory" must be wrong. Just as we had to think carefully about the details of the experiment, we also have to look closely at what went into the "theory". The reason "theory" is in quotation marks is that it actually has several parts to it, any one of which could be the culprit.

> There are three separate parts to the theory. See if you can identify them before reading further.

10.5.3.1 Look Closely at the Theory

We have already mentioned several times the most prominent parts of the theory, namely, the conservation of momentum and energy. Being conservative, physicists are very loathe to give up either one of these. After all, there is a great deal of evidence that they are correct, and remember, each is based on an important symmetry principle. In fact, when the beta decay spectrum was first measured in about 1930, it was suggested that perhaps these conservation laws did not hold for nuclear processes. But as we shall see below, there is a much more mundane explanation.

The other part of the theory is the model: the picture we have in our mind of beta decay where the neutron explodes into a proton and electron, as depicted in Figure 10.3. Since a model, as we discussed back in Chapter 2, is just a picture we make up in our mind to help us visualize something we cannot directly perceive, it is certainly reasonable to question this first. The question then is: How can we modify the model to make its prediction agree with the experimental results? We might expect to get some hints by looking at those results.

10.5.3.2 Look Closely at the Experimental Results

Figure 10.5 is a plot of the number of electrons emitted as a function of the emitted energy. The first thing it tells us is that there is a spectrum of energies, not just one unique energy. Also, just about all the electrons are being emitted at energies lower than the predicted energy (note that both of these contradict the prediction). It appears that a very small number come off at the predicted energy but there are none above that energy. (Did you see any of this when you were asked to think of the message of Figure 10.5?)

10.5.3.3 A Possible Explanation

How does this help us? Well, since the electrons have less than the expected energy, there is a loss of energy. We have already discussed the possibility of loss of energy in passing through the material in which the nuclei are embedded. How else can the electron come off with less energy than expected? One possibility is that there is another particle being emitted, an X particle, which can then carry away some of the energy. This is essentailly a guess, but let us see where it leads us.

First, how does this help us undersrtand why the electron is emitted with a range of energies? Because there are now three particles coming off that can all share the energy and momentum. If only the electron and proton are emitted, then by conservation of momentum:

$$p_e + p_p = 0 \tag{10.5}$$

or

$$p_e = -p_p \tag{10.6}$$

The electron and proton must have the same value of momentum. But if three particles are emitted, then we have:

$$p_e + p_p + p_X = 0 \tag{10.7}$$

or

$$(p_e + p_X) = -p_p \tag{10.8}$$

We see p_e can now have a continuous range of values as long as it and p_X add up to $-p_p$. The only thing that actually limits the values of p_e is the conservation of energy, as we shall see below.

But if the X particle exists, where are its properties? We can be helped in this by writing the decay equation for the neutron decay with the X particle:

$$N \rightarrow p + e + X$$

By conservation of charge, the X particle must be electrically neutral. We see this by explicitly writing the charge in the equation:

$$n^0 \rightarrow p^{+1} + e^{-1} + X^0$$

We can also say something about the mass of the X particle by writing the conservation of energy equation, modifying Equation 10.3 by adding on the mass-energy of X:

$$938 + 0.51 + m_X + KE = 939 \tag{10.9}$$

where, remember, KE is the kinetic energy of all the emitted particles. Again, we can ignore the KE of the proton, as we did before.

In order to see where to go from here, let us take a closer look at the experimental distribution in Figure 10.5. It tells us that at low energies, near zero, there are relatively few electrons emitted. As the energy increases up to a value about half the predicted energy, the number of emitted electrons also increases. Above about half the predicted energy, the number of electrons decreases, going to zero at or near the predicted value of energy. What this means is that, on average, the electron and X share the energy equally. That is why the maximum number of electrons (and X particles) have about half the initially predicted energy. But when the electron has very little energy, near zero, the X particle has its maximum allowable energy. And when the electron has its maximum allowable energy, at 0.49 MeV, the X particle has zero energy. In other words, the total KE of the electron and X particle is 0.49 Mev, and they share it as described by Figure 10.5. So, in Equation 10.9, we can put in 0.49 for KE and write:

$$938 + 0.51 + m_X + 0.49 = 939 \tag{10.10}$$

or

$$m_X = 939 - (938 + 0.51 + 0.49) = 0 \tag{10.11}$$

m_X would indeed be zero if we could measure that some electrons do have a KE of exactly 0.49 MeV. But it is extremely difficult to measure it exactly. The important point here is that the mass of the X particle is much smaller than the electron mass and could be zero, or very close.

So, does this elusive particle with no charge and very little mass really exist? For about 30 years, it was not at all clear. But, now we know the answer is yes! It was named by the great Italian physicist Enrico Fermi as the neutrino, which in Italian means "little neutral one." The neutrino is indeed very elusive. The beta decay spectrum shown in Figure 10.5 was first measured about 1930. While the neutrino was hypothesized shortly afterward to explain the measurement, it was not observed until 1958. It is extremely difficult to detect because it hardly interacts with anything. With no electric charge, it only interacts through the weak nuclear force that has a strength of about 10^{-7} times that of the electromagnetic force. A neutrino can go through the earth without interacting. The vast majority of neutrinos in the universe are born in the burning core of stars. They can go through an entire star without interacting. In fact, because they do not interact, they escape the star depositing only very little of their energy in the star. Neutrino energy loss is the main mechanism for stellar cooling. If neutrinos did leave some fraction of their energy in a star, like our sun, for instance, then the sun would have burned up a long time ago and we would not be here reading about neutrinos.

One last update to our story. It was assumed, even though it could not be measured exactly, that the neutrino mass was indeed zero. But recent experiments have shown that the neutrino has a nonzero mass. Interestingly, while we know it is nonzero, we have still not been able to measure its true value. We do know that it is very small, probably less than 1 eV (.000001 Mev).

10.6 END-OF-CHAPTER GUIDE TO KEY IDEAS

- What is the nucleus made of?
- What holds it together?
- Why are neutrons needed?
- What are the three different types of nuclear decay? What is the nature of the particle emitted in each case?
- Why do some nuclei undergo radioactive decay?
- What is the definition of *half-life*?
- Can you describe carbon dating?
- What assumptions are made in determining the age of something by carbon dating? How are they verified?
- Why was it necessary to propose the existence of the neutrino? There are several parts to this story—can you tell the full story?
- What are the properties of the neutrino?

QUESTIONS/PROBLEMS

1. What is unique about the hydrogen nucleus?
2. List all of the properties of the proton. The neutron.
3. How are protons and neutrons similar?
4. How are protons and neutrons different?
5. Can gravity hold the nucleus together? Explain your answer.
6. Can the electromagnetic force hold the nucleus together? Explain your answer.
7. Define atomic number, neutron number, and atomic weight. Give the symbols for each.
8. Why is the symbol A known as the atomic weight?
9. Define isotope.
10. What is the purpose of neutrons in the nucleus?
11. Mercury, symbol Hg, has an atomic weight of 200. The radius of the nucleus of Hg is 6×10^{-15} m.
    a. If density is defined as mass/volume, find the density of the Hg nucleus in units of kg/m^3. Assume the nucleus is spherical in shape.
    b. The density of the element mercury is 13,600 kg/m^3. Given the answer you found in (a), does this make sense?
12. What is the purpose of nuclear decays?
13. How are nuclear decays similar to the radiation emitted in atomic spectra? How are they different?
14. Define the three different forms of nuclear decay.
15. Which type of nuclear decay rarely occurs in the lighter nuclei?
16. Which type of nuclear decay can cause the worst biological damage? Why?
17. What is the basic process that occurs in β-decay?
18. In β-decay, which properties of the electron and proton are the same? Which are different? Why?
19. In which nuclear decays is the atomic weight the same before and after the decay? In which is the atomic number the same?

20. Which nuclear decay is most like the radiation that produces atomic spectra? How is it similar and how is it different?
21. Consider a nucleus with 80 protons and 120 neutrons.
    a. What element is this?
    b. Write the decay equation (including appropriate element symbol, atomic number, and atomic weight) if this nucleus undergoes:
        i. α-decay
        ii. β-decay
        iii. γ-decay
22. Define half-life.
23. If a radioactive sample originally has N nuclei, how many are left after three half-lives?
24. The half-life of free neutrons is 15 min. If you start off with 1,200 neutrons, how many are left after 30 min? How many have decayed?
25. In any sample of carbon, there is a minute amount of ^{14}C. How do we measure the amount of ^{14}C in a sample of carbon?
26. How is ^{14}C produced?
27. A wooden carving has 1 g of carbon in it and the ^{14}C decay rate has been measured to be 8 decays/min. How old is the carving?
28. A wooden carving has ½ g of carbon in it and the ^{14}C decay rate has been measured to be 4 decays/min. How old is the carving?
29. A wooden carving has ½ g of carbon in it and the ^{14}C decay rate has been measured to be 8 decays/min. How old is the carving?
30. What conservation laws are used to predict the kinetic energy of the electron in β-decay?
31. Why isn't the proton observed in β-decay?
32. In β-decay, how does the experimental result differ from the theoretical prediction?
33. Explain one way in which the results in the β-decay experiment could be wrong. How could that be tested?
34. List all the things you can as to what the graph of the experimental results in Figure 10.5 tell us about the kinetic energies of the emitted electrons.
35. Discuss how the neutrino hypothesis explains the experimental curve shown in Figure 10.5.
36. Explain why the mass of the neutrino must be very close to zero.
37. List all of the properties of the neutrino.
38. If the mass of the neutrino was 0.2 MeV, draw the graph of the distribution of electron kinetic energies in β-decay. Be sure to put in relevant numbers.
39. Why is the neutrino so important for our existence?

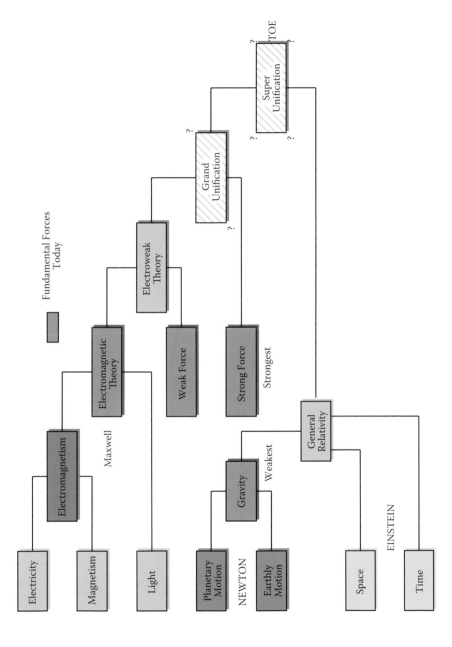

FIGURE 4.1 The unification of forces.

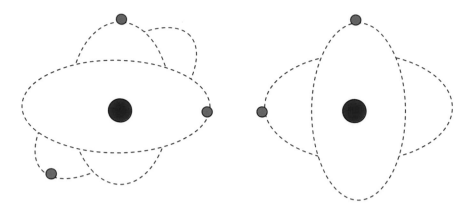

FIGURE 6.3   Two atoms (hand and table) close to each other. The dashed ellipses represent the electron (blue) orbits that circle the nucleus (red circles). This is not to scale since the size of the electron orbits is about 100,000 times the size of the nucleus.

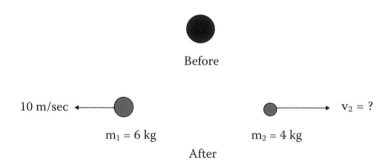

FIGURE 8.1   Explosion of a 10 kg bomb initially at rest.

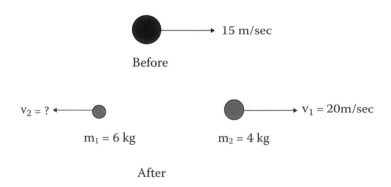

FIGURE 8.2   A 10 kg bomb moving with a speed of 15 m/s before exploding.

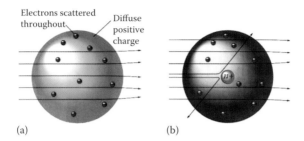

(a)   (b)

FIGURE 9.1 (a) If the plum pudding model were correct, alpha particles would pass through the atom with little or no deflection. (b) What Rutherford observed was that some alpha particles were deflected through large angles, indicating the presence of a massive nucleus. (From Ted Ankara; library.tedankara.k12.tr/chemistry/vol3/vol3.html)

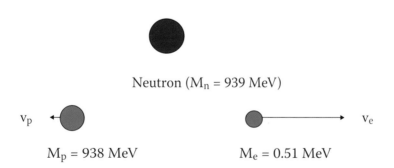

Neutron (M_n = 939 MeV)

M_p = 938 MeV M_e = 0.51 MeV

FIGURE 10.3 A picture of beta decay where a neutron decays (explodes) into a proton and an electron.

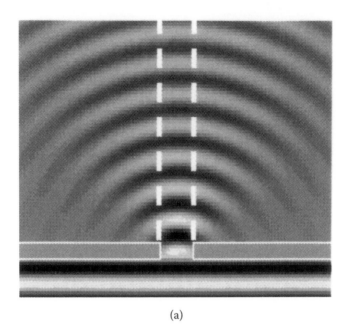

(a)

(b)

FIGURE 11.5 Diffraction of waves through an opening. In the upper part of the figure (a), the opening is small or comparable to the wavelength, and a good deal of the wave is bent behind the barrier. On the lower part (b), the opening is a good deal larger than the wavelength, and most of the wave goes forward, more like particles would behave. (From Chiu-King Ng; www.ngsir.netforms.com)

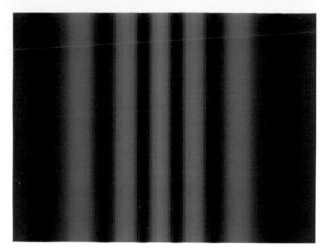

FIGURE 11.7 The result of shining light on two slits showing a typical intereference pattern for waves. (From M. Goldmen; www.colorado.edu/physics/2000/images/two_slit_logo.jpg)

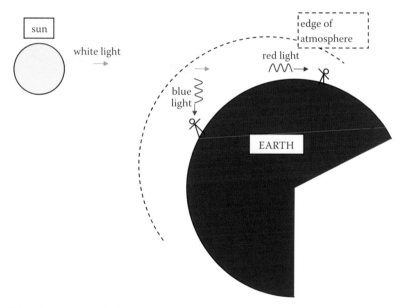

FIGURE 11.8 Why the sky is blue and sunsets are red.

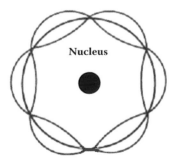

FIGURE 13.1 de Broglie wavelengths in the circular electron orbit in the atom. This particular configuration corresponds to three integral number of wavelengths (n = 3).

FIGURE 14.2 In our most modern view of forces, two electrons exchange a photon (γ), the force carrier associated with the electromagnetic force.

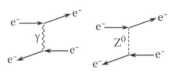

FIGURE 14.3 Feynman diagrams depicting the force between two electrons due to the exchange of a photon (electromagnetic force) and the exchange of a Z^0. When the kinetic energies of the electrons are much greater than the mass-energy of the Z^0 (100 GeV), the force due to either exchange becomes the same. For energies below 100 GeV, the two forces do not behave the same.

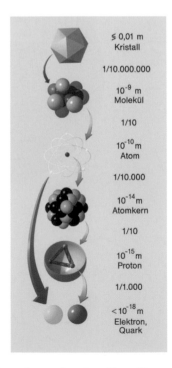

FIGURE 14.4 The ultimate makeup of matter. (From Desy; www.zms.desy.de/e548/e550/e6943/e83/imageobject148/kristall_quark_hr_ger.jpg)

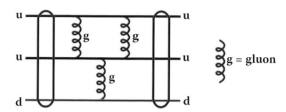

FIGURE 14.6   A proton is made up of three quarks: two up quarks and one down quark. The quarks are held together by the exchange of the strong force carriers, the gluons.

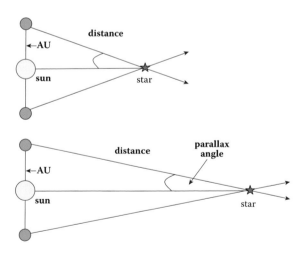

FIGURE 15.1   Two examples of parallax measurements: one for a relatively nearby star and one for a star farther away. The parallax angle is smaller the farther the star.

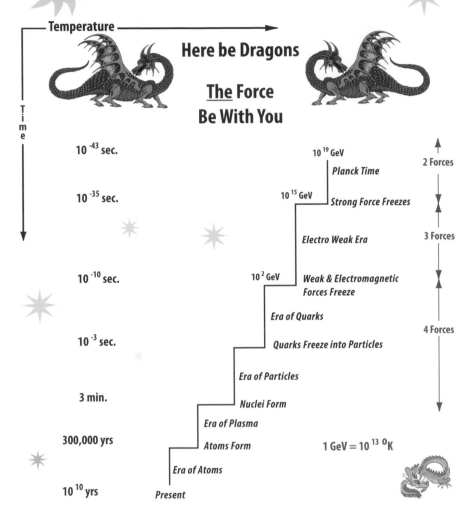

FIGURE 15.2 The evolution of the universe from the big bang to the present. (Adapted from *Moment of Creation* by James Trefil.)

11 The Nature of Light

11.1 INTRODUCTION

At the beginning of the twentieth century, two great revolutions in physics occurred that changed the way physicists viewed the world. One is known as quantum mechanics and the other is Einstein's theory of relativity. Interestingly, they both were initially concerned with the nature of light. Quantum mechanics began with an attempt to understand how light interacted with matter, while an integral part of relativity has to do with the speed of light. Therefore, it is fitting that before we discuss these two great theories, we take some time to discuss light and its properties.

There has been a great debate since at least Newton's time as to whether light is a wave or particle phenomenon. Newton believed light was made up of particles. But in 1800, Thomas Young performed an experiment that could only be explained if light was a wave phenomenon. Then in 1865, Maxwell showed theoretically that light was an electromagnetic wave.

Waves and particles appear to be very different phenomena. In order to understand their differences, we have to consider the properties of each.

11.2 PROPERTIES OF PARTICLES

The common image of a particle is a tiny ball of mass like a BB. In other words, it is well localized in space and is solid. No two particles can be in the same place at the same time. For our purposes here, this describes the basic properties of particles. As we will see below, waves require a longer discussion.

11.3 PROPERTIES OF WAVES

You are most likely familiar with many different types of waves: water waves, waves on a string, people waves (in a stadium), and sound waves in air or water, to name a few. What do all of these have in common?

> Can you think of a good general definition of a wave? Try to do this before reading further.

Notice we have described these waves in terms of what is waving, in other words, the *medium* for that wave. In fact, a wave is a *disturbance of a medium*. For example, consider a smooth lake into which a pebble is dropped. The pebble causes ripples on the surface of the lake that move outward from the location of the dropped pebble. These ripples, or disturbances, are the water waves. We see, just from this description, that waves, unlike particles, are not well localized but tend to propogate throughout the medium.

Before going further, we should define several important quantities that are necessary to fully describe waves. Figure 11.1 depicts an idealized wave.

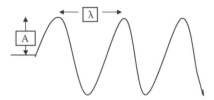

FIGURE 11.1   A sinusoidal wave that repeats its pattern. The distance between identical parts of the pattern is known as the wavelength, λ. The maximum disturbance as measured from the undisturbed position to the crest or trough position is the amplitude, A.

11.3.1   WAVE VOCABULARY

- **Wavelength** (symbol λ): The distance between identical parts of the wave pattern (e.g., the distance between crests or troughs). For ocean waves it could be 5 to 10 ft; for a typical sound wave it is about 1 ft; and for green light, the color in the middle of visible light, the wavelength is 5×10^{-5} cm.
- **Period** (symbol T): The time it takes one wavelength to go by. If one looks at one point along the wave, it is also the time for the wave to make one complete oscillation. The period for green light is an extremely short time, 2×10^{-15} s.
- **Frequency** (symbol f): The number of oscillations the wave disturbance makes in a certain time interval—usually in 1 s. For instance, if one counts eight oscillations in 2 s, the frequency would be 4 oscillations/s. Often, frequency is given in units of Hertz, or Hz, which is 1 oscillation/s: 4 oscillation/s $=$ 4 Hz. From our definitions of period and frequency, there is a simple relation between the two:

$$T = 1/f$$

  For sound waves, our ears are sensitive to frequencies from about 100 to 10,000 Hz. The frequency for green light is 5×10^{14} Hz.
- **Wave speed** (symbol v): The speed the wave moves with respect to its medium. In fact, this speed is determined entirely by the properties of the medium. For example, a wave on a string, such as a violin, guitar, or piano string, is determined by two properties of the string: the mass per unit length (also known as the linear density) and the tension in the string. Those of you who play a violin or guitar know that the different strings are of different thickness (hence, different linear density), and you can tune the string by changing its tension.

The important thing to remember here is that the medium plays a crucial role for any wave. Without a medium there is nothing to be disturbed, and hence no wave. It is the properties of the particular medium that determines the speed of that particular wave.

- **What determines λ?**
  There is an important relation between v, f, and λ, which in fact determines the value of λ since f is determined by the source of the wave. It is given by the equation

$$\lambda = v/f = vT \tag{11.1}$$

- **Wave amplitude** (symbol A): The maximum size of the disturbance as measured from the undisturbed position to the crest or trough.
- **Wave types:** There are two important directions of motion associated with any wave. One is the direction of motion of the wave, also called the propagation direction. The other is the direction of motion of the medium. For instance, if you drop a pebble in water, the wave propagates horizontally along the surface of the water and radially outward from where the pebble entered the water. But the motion of the disturbance (the water) is up and down at right angles to the propagation direction.

  Transverse wave: A wave as just described above is called a transverse wave. Generally, it is a wave where the direction of oscillation of the medium is at right angles to the propagation direction. Waves on a string are also transverse.

  Longitudinal wave: A wave where the direction of oscillation of the medium is in the same direction as the propagation direction. Sound waves are one of the most common examples of longitudinal waves. Another example is a wave on a spring produced if part of the spring is either compressed or stretched.

- **Polarization:** This is only defined for transverse waves. For such a wave, the medium can oscillate in any direction in the plane perpendicular to the propagation direction. If there is a preferred direction of oscillation, we say the wave is polarized. For example, if a wave is coming toward you out of the page, it might oscillate only up and down (vertical polarization). Or it might oscillate left and right (horizontal polarization). It could also oscillate from the upper left corner to the lower right corner (diagonal polarization). Also, it is possible to have no preferred direction of oscillation. In that case, the wave is unpolarized. Since a longitudinal wave has only one possible direction of oscillation, polarization only applies to transverse waves.
- **Superposition:** The ability of waves to be in the same place at the same time and have their disturbances add. Figure 11.2 shows two waves of different wavelength and amplitudes. The bottom wave shows the result of adding the two disturbances from the two original waves.

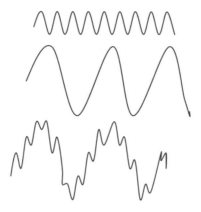

FIGURE 11.2   Superposition of two waves with different wavelengths and amplitudes.

- **Interference:** This is a special case of superposition when both waves have the same wavelegth (i.e., the same, identical shapes). There are two important cases, as shown in Figure 11.3. Constructive interference occurs when the two waves combine crest to crest. Here we say the waves are in phase.

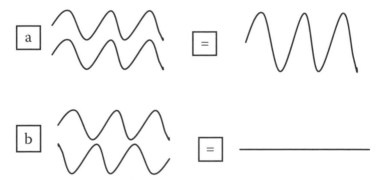

FIGURE 11.3   (a) Constructive interference. (b) Destructive interference.

The resultant wave has an amplitude twice as large if the two waves have equal amplitude. Destructive interference occurs when the waves are crest to trough. The waves are out of phase, and the resultant amplitude is zero if the two combining waves have equal amplitude. When there is a continuous source of two waves as depicted in Figure 11.4, there is what is usually called an interference pattern, where there are definite regions of constructive and destructive interference. The constructive interference regions are where the resultant wave amplitude is a maximum, and the destructive interference regions are where the resultant wave amplitude is minimum. (If the two waves are of equal amplitude, then this minimum is zero.) The shape of the pattern (i.e., the angles between minima and maxima) is determined by the space between the two sources and the wavelength of the waves. The difference between a maximum and a minumum corresponds to a difference in path length of travel of half a wavelength, or a difference in time of arrival of half a period for the two waves. Interference can provide extremely accurate measurements of distance or time differences if the waves used have very small wavelengths or periods. We shall see an application of this in the next chapter.

Only waves can exhibit interference since two particles cannot be in the same place at the same time. Thus, interference is one of the main tests as to whether something is a wave or particle phenomenon.

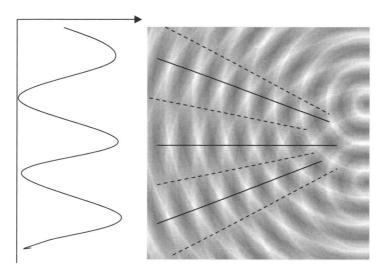

FIGURE 11.4   Interference pattern of water waves. Two sticks continually dipped in the water act as sources of circular waves. Constructive interference regions are seen as the bright spots indicated by the solid lines in the picture. Destructive interference is seen as the darker regions, as indicated by the dashed lines. (Adapted from the Open Door Web Site, www.saburchill.com/physics/chapters2/0007.html) The plot to the left shows what a graph of the intensity pattern would look like. The maximum intensity corresponds to a position of constructive interference, and zero intensity corresponds to destructive interference.

- **Diffraction:** The ability of waves to bend around corners or obstacles. Because waves can diffract, it is possible to observe a wave behind an object. For instance, you can be in a room with an open door and still hear someone who is outside and around the corner. In order for that to happen, the sound waves had to be able to bend around the corner to get to your ears. The amount a wave is bent depends on both the size of the opening and the wavelength, as can be seen in Figure 11.5. Figure 11.6 depicts diffraction around an obstacle and through an opening showing that they are both due to the same effect. In both cases, the diffraction of the wave is easily seen.

    A very important fact to realize here is that if we want to observe an object using waves of any type, then diffraction limits our ability to observe that object if it is small compared to the wavelength of the wave. In order to observe something with any type of signal, the object must change the nature of the signal so we know the signal has been disturbed by the object.

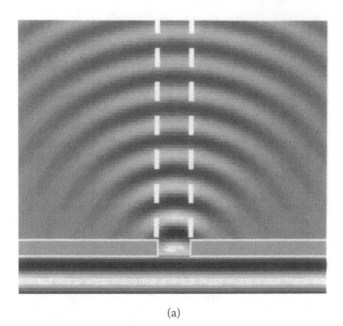

(a)

(b)

FIGURE 11.5 **(See color insert following page 102.**) Diffraction of waves through an opening. In the upper part of the figure (a), the opening is small or comparable to the wavelength, and a good deal of the wave is bent behind the barrier. On the lower part (b), the opening is a good deal larger than the wavelength, and most of the wave goes forward, more like particles would behave. (From Chiu-King Ng; www.ngsir.netforms.com)

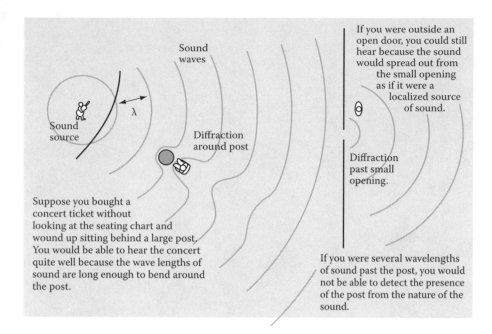

Sound waves

Sound source

λ

Diffraction around post

If you were outside an open door, you could still hear because the sound would spread out from the small opening as if it were a localized source of sound.

θ

Diffraction past small opening.

Suppose you bought a concert ticket without looking at the seating chart and wound up sitting behind a large post. You would be able to hear the concert quite well because the wave lengths of sound are long enough to bend around the post.

If you were several wavelengths of sound past the post, you would not be able to detect the presence of the post from the nature of the sound.

FIGURE 11.6 Diffraction of water waves around an obstacle. Notice that directly behind the obstacle the wave is blocked out, or shadowed. A little further on, the wave reforms. If this were light, it would be a bright spot behind the obstacle. (From Rod Nave; www.hyperphysics.phys-astr.gsu.edu/hbase/sound/imagesou/difr.gif)

If there is too much diffraction, then the wave (the signal in this case) a short distance from the obstacle appears as it would if nothing were there.

11.4 IS LIGHT MADE UP OF WAVES OR PARTICLES?

Now that we have defined the properties above, we are ready to discuss whether light is made up of particles or waves.

One of the primary tests for waves is to see whether the phenomenon shows interference. To do this, we can set up what is called a double slit, where light is caused to be incident on two narrow slits. If light is a wave, the slits act as sources of circular waves, the same as the dippers in Figure 11.4. The two circular waves would then interfere, giving an intenstiy pattern as shown in the figure. If, for instance, a green laser was used as the source of light, then multiple green (bright) spots would be seen on a screen placed after the slits. On the other hand, if light was made of particles, then only two bright spots, directly in line with the slits, would be seen. Figure 11.7 shows the results of performing the two-slit experiment. Clearly there are multiple bright spots showing light must be a wave phenomenon.

In addition to interference, light also exhibits diffraction. You can see diffraction for yourself by putting two adjacent fingers close together so that there is a small spacing between them and looking at a light source through the spacing. If your fingers are not too close together, you will just see the light, but if you make the spacing

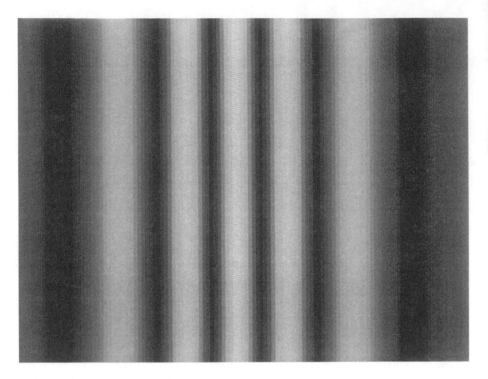

FIGURE 11.7   (**See color insert following page 102**.) The result of shining light on two slits showing a typical intereference pattern for waves. (From M. Goldmen; www.colorado.edu/physics/2000/images/two_slit_logo.jpg)

small enough, you will see dark lines in the spacing between your fingers. This is due to the two waves that have been diffracted around the two edges of your fingers interfering with each other.

We might also ask whether light is a transverse or longitudinal wave. We can test this by using a polarizer, which allows only tranverse waves oscillating in one direction (the axis of the polarizer) to be transmitted. If light is a transverse wave and the source of the light produces completely polarized light, then a polarizer with its axis at right angles to the polarization direction of the light will block out all the light. On the other hand, if the light source produces completely unpolarized light, so there is no preferred direction of oscillation, then the polarizer would not block out all of the light. But if we use two polarizers, we can do a definitive test even if the light is initially unpolarized. If we allow the light to pass through the first polarizer, then, if it is transverse, the light passing through the first polarizer will be polarized along the polarizer axis direction. If this light is now incident on the second polarizer, and that polarizer is oriented at right angles to the first one, then the light will not be transmitted through this second polarizer. If light is a longitudinal wave, then polarizers will have no effect on it. Hopefully, you will see this demonstrated in class, but two polarizers crossed at right angles always extinguishes light. Light is a transverse wave.

There is also a very dramatic demonstration using three polarizers that shows that light is a vector transverse wave. In other words, whatever is waving has vector properties of both magnitude and direction. We will get into the discussion of what is waving in the next chapter. Possibly you will be able to see the three-polarizer demonstation in class.

At this point, it would appear from the unmistakeable evidence of interference, diffraction, and polarization that light is certainly a wave phenomena. You should be warned that in Chapter 13, we will discuss the unmistakeable evidence that light is a particle phenomena.

11.5 BACK TO DIFFRACTION

Above we dicussed the fact that it is diffraction that limits our ability to observe small objects clearly with waves. Now that we know light is a wave, this has important consequences for our being able to see small objects. Remember, the smaller the object is compared to the wavelength of light, the more the light is bent around the object. A typical wavelength for light is about 5×10^{-5} cm. That means objects with sizes comparable to that will look rather fuzzy to us. For objects much smaller than that, we will not be able to see them at all. An atom has dimensions of the order of 10^{-8} cm, much smaller than the wavelength of light. Given this discussion, it is clear that it is impossible to see an atom with our eyes. This has nothing to do with the possibility of inventing a better optical microscope. It is a limit imposed by nature.

On the other hand, there are certainly many small objects that we can see quite clearly in a microscope or with the unaided eye. We can consider ourselves very lucky that our eyes are not sensitive to radio waves, which have wavelengths of meters or longer. If this were the case, we would all look very fuzzy, at best, to each other.

11.6 WHY THE SKY IS BLUE

This is one of the most common questions children and adults ask. Since this chapter is about the nature of light, it seems appropriate to give the explanation here.

These are the relevant facts:

- The light in the sky is coming from the sun.
- When we are looking at the sky, we are looking away from the sun.
- Thus, we are looking at scattered light, i.e., light that has changed its direction due to interactions with the molecules in the air.
- The light from the sun is made up of many colors. You see this when you see a rainbow, which is caused by water droplets in the air acting as prisms.
- The scattering process is such that high-frequency (blue) light is scattered more than low-frequency (red) light.
- Since more of the high-frequency blue light is scattered into your eyes, that part of the sky, where the light is coming from, looks blue.

There is also a secondary result of the sky being blue. If the higher-frequency light is scattered out then what is left is the lower-frequency, red, light. Thus, if you look at light that has had a lot of the blue light scattered out—light that has

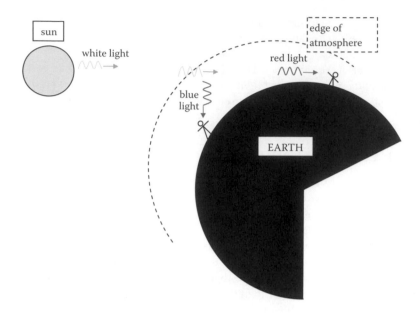

FIGURE 11.8 (**See color insert following page 102.**) Why the sky is blue and sunsets are red.

gone through a lot of atmosphere—it will look red. When does this occur? At sunsets when the light has traveled through the most atmosphere. Thus, sunsets are red because the sky is blue!

11.7 END-OF-CHAPTER GUIDE TO KEY IDEAS

- What is a particle?
- What is a wave?
- Can you define the ten different wave vocabularly terms?
- Is light made up of waves or particles? What is the evidence?
- What limits our ability to see small objects? Why?
- Why is the sky blue?
- Why are sunsets red?

QUESTIONS/PROBLEMS

1. What are the different models for what light is made of?
2. List all the properties of particles.
3. Define a wave. List several types of waves you know of.
4. List six properties necessary to describe a particular wave. Three of these properties are related. What are the relations between these three?
5. Define *transverse wave* and *longitudinal wave*. Give at least one example of each.
6. Define *polarization*.

7. Define *superposition*.
8. How is interference different from superposition?
9. Define *constructive interference* and *destructive interference*?
10. Why can't particles interfere?
11. Define *diffraction*.
12. What factors determine whether diffraction is easily observable?
13. Under what conditions can an object be observed using a wave?
14. What is the evidence for light being a wave phenomenon?
15. Given the best technology possible, can we observe an atom with visible light? Justify your answer.
16. What would we experience if our eyes were sensitive to radio waves instead of visible light?
17. Why is the sky blue?
18. Why are sunsets red?
19. What would be the color of the sky if there were no atmosphere?
20. If the atmosphere were less dense (fewer particles in it), would sunsets be more red or less red? Justify your answer.

12 The Theory of Relativity

12.1 INTRODUCTION

You may have already heard some things about relativity. Perhaps you have heard that it says strange things about time and length, or it does not make sense (i.e., it does violence to your intuition), or only twelve people in the world understand it. The first two statements are true, but the last is certainly not. While some of the ideas of relativity will indeed initially do violence to your everyday intuition, they also follow logically. Hopefully by the end of this chapter, you will have a fairly good understanding of relativity.

Basically it has to do with what two people observe when they are moving relative to each other. Thinking about things like this, in the mind of an Einstein, has led to a new and profound understanding of the universe, including how stars shine. It has led to new technologies such as nuclear energy, which has affected our lives through nuclear power generation and geopolitics. As we shall see, it is based on a symmetry principle.

12.2 FRAMES OF REFERENCE AND RELATIVE SPEEDS

Consider the following question: How fast are you moving at the instant you are reading this? What is your answer? You might say, "Well, I'm sitting in my chair, so I am not moving at all." But what if you are in your friend's car, moving down the road at 50 mph? Would you say 50 mph, or would you say, "Well, with respect to the car I am not moving"? Which is the correct answer in this case? Here, the answer is ambiguous because the question itself is ambiguous. The question "How fast am I moving?" has no meaning. It must be referenced to something. "How fast am I moving with respect to the ground?" or " How fast am I moving with repect to the car?" are proper questions because they have well-defined and understood answers. To make the point, the initial question about how fast you are moving while reading the book has, technically, an infinite number of answers. To give just a few: relative to your room, the answer is zero; relative to the sun, the answer is about 18 miles/s; and relative to the center of the galaxy, it is about 150 miles/s. There is no single, or *absolute*, answer. Or another way of stating this is there is no *absolute frame of reference* that all speeds can be referenced to (i.e., there is no place in the universe that we can say is *at rest*). We can only measure speeds with respect to some *frame of reference*.

If we have two observers moving relative to each other, how do we relate the speed of some object that they both measure? As one simple example, in your own reference frame you are at rest, but to someone who sees your reference frame move at 20 mph, that observer will say you have a speed of 20 mph. We want to think about the more general case where there is an object that is moving as measured in both reference frames. What speed does each observer measure? We will write down a general equation, but most of you know how to do this intuitively.

For example, consider being on a railroad train car moving at 60 mph relative to the ground. You throw a ball at 20 mph in the same direction the train is moving. What does someone on the ground measure for the speed of the same ball? Hopefully you see, intuitively, that the answer is 80 mph (20 + 60). On the other hand, if you throw the ball in the opposite direction to the train's speed, the ground observer will measure a speed of 40 mph (60 − 20). It is important to note that the two observers are measuring the speed of the same object, but get different answers due to their relative motion. Now let us obtain a general equation that should be true for any speed, not just 60 and 20 mph.

Look at Figure 12.1.

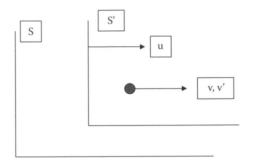

FIGURE 12.1 Reference frames S and S′ with S′ moving with speed u to the right relative to S. The ball has a speed v as measured by an observer in S, and v′ measured in S′.

Using our example above, the train's speed of 60 mph is now the relative speed of the two reference frames, u. Twenty miles per hour is the speed v′, and 80 mph is the speed v. So, we can write generally,

$$v = v' + u \tag{12.1}$$

This is just the equation Galileo had written some 400 years ago and is, in fact, known as the Galilean transformation of speeds. While this looks more formal, hopefully you see that it is just saying what your intuition tells you.

EXAMPLES

1. Bob and Mary are in two airplanes that pass each other. Bob throws a ball in his plane at a speed of 50 mph, which Mary measures as 250 mph. What is the relative speed of the two planes?

   Here we are asked to find the speed u given v and v′, where v is 250 mph and v′ is 50 mph. Solving Equation 12.1 for u, we have

   $$u = v - v' = 250 - 50 = 200 \text{ mph}$$

   We can also tell that u, v, and v′ are all in the same direction either by intuition or by the fact that they all have the same sign. For future reference, let us call this direction "to the right."

2. Bob and Mary are in the same planes with the same relative speed. Mary measures the speed of the ball that Bob throws as being 150 mph. What is the direction and speed at which Bob throws the ball in his plane?

Here we want to find v' given that u is 200 mph and v is 150 mph.

$$v' = v - u = 150 - 200 = -50 \text{ mph}$$

The minus sign tells us that Bob has thrown the ball in the opposite direction that his plane is moving relative to Mary's.

3. To show there is nothing special about either Bob's or Mary's reference frames (planes), let us have Mary throw a ball at 50 mph in the same direction her plane is moving. What does Bob measure? We can do this in two different ways.

First, let us keep "to the right," the direction of motion of Bob's plane as positive. Then u = −200 and v' = −50, so we get

$$v = v' + u = -50 - 200 = -250 \text{ mph}$$

In other words, Bob sees the ball Mary has thrown as moving to the left (the same direction he sees Mary's plane moving in).

Or, we can define "to the left" as positive. Then both u and v' are in the positive direction, and we have

$$v = v' + u = +50 + (+200) = 250 \text{ mph}$$

Now the positive sign means to Bob's left.

12.3 GALILEAN RELATIVITY

We see that the concern about relative motion goes back to at least Galileo, some 300 years before Einstein. The principle of Galilean relativity is easily stated as:

All inertial reference frames are equivalent.

This simple-sounding statement has important consequences, so let us make sure we understand all the relevant terms.

1. A reference frame is the location from which some observer is making whatever measurements he or she is making. It could be the room he or she is sitting in, or a car, train, plane, or rocket ship in which he or she is riding.

2. An inertial reference frame is any reference frame that is moving with constant *velocity*. Note it does not say *speed*, but *velocity*. In other words, the reference frame cannot be changing its speed or direction. It is a nonaccelerated frame. The relativity theory we will be discussing is called special relativity because it has to do with inertial reference frames only. About 10 years after Einstein formulated special relativity, he introduced general relativity, which broadened considerations to include accelerated frames also. Thinking about noninertial reference frames led Einstein to formulate a new theory of gravity that, in principle, differs from Newtonian gravity.

For most situations, general relativity agrees with Newtonian predictions. But for extreme situations, such a black holes or the universe as a whole, evidence shows that Einstein's formulation is correct.

3. Equivalent means that no inertial reference frame is any better or worse than any other in the sense that there is no way of detecting your *absolute* motion in any inertial frame. In other words, you cannot tell whether you are *at rest* or in motion. The idea of at rest only has meaning in a relative sense, but not in any absolute sense. You can claim you are at rest with respect to someone else, for instance, someone sitting next to you. But the absolute statement "I am at rest" has no meaning since there is no way of detecting your motion or lack of it. This is one of the hardest ideas to come to grips with in relativity, but it is indeed true. In fact, most of you reading this have probably experienced this lack of ability to know whether you are moving or not. Have you ever been on a train in a station when the train next to you has started but you have the feeling that your train is the one that is moving? Or your train has gently begun to move, but you feel the train next to you has started? Another example is being in an airplane at cruising speed. You have no senstation of moving at 600 mph if you do not look out the window. This last statement is important. Of course, when you are in a car going down the road, you know you are moving since you look out the window and see the trees and buildings moving by. You have your usual cues from everyday experience. But if you were in a rocket ship, in deep outer space, you would not have such cues. If you then looked out of your rocket window, you could not detect whether you were in motion or not. If you saw another rocket move by, you would not know whether that rocket was moving to your left, for instance, and your rocket was at rest. Or you were moving to its right and it was at rest. Or both rocket ships were moving. There would be no measurement you could make to tell your absolute motion. The whole notion of absolute rest or motion has no meaning!

As you can see, we have devoted quite a bit of space to that "simple" statement of Galilean relativity. There is one other important thing to point out. Galilean relativity is based on a symmetry principle. Remember our definition of *symmetry*: something is symmetric if we can find an operation such that after performing that operation, we cannot tell that anything has been done.

> Do you see how the statement of Galilean relativity is a symmetry principle? Before going on, try to think about this.

What is the operation we want to consider in this case? It is transferring an observer (you, for example) from one inertial reference frame (e.g., a rocket in outer space) to another. Let us assume that you are asleep, so you do not know whether anything has been done. When you wake up, is there any way of knowing whether you have been transferred to another inertial rocket frame? (We assume, of course, that the rocket ships are identical, so you cannot tell, for instance, by noting the furniture is different.) Since there is no way of detecting your absolute motion, there is no measurement or observation you can make that will allow you to know whether any transfer has taken place. So, from our definition, Galilean relativity is based on a symmetry principle.

You may think that it is just the symmetry of space translation, which we have discussed previously. But in fact, it is the symmetry of velocity translation, which is different from just moving you from one location to another.

12.4 MAXWELL AND THE ETHER

In order to fully appreciate the ideas associated with relativity, we will have to go back in history and discuss some ideas that were believed to be true at the time but are now known to be wrong. As you will see, this will help us in our final understanding of the theory.

As was discussed above, according to Galilean relativity, it is impossible to detect absolute motion. Indeed, this was believed from Galileo's time to the year 1865. In that year, James Clerk Maxwell, who is considered the father of the electromagnet theory, discovered something that implied that there was a universal, absolute frame of reference. In other words, absolute speeds could be measured from this reference frame.

Maxwell did the final unification of electricity and magnetism. When he combined Faraday's law with his modified version of Ampere's law, he showed that light was an electromagnetic wave with a speed of 186,000 miles/s or 3×10^8 m/s. What is important here is the fact that light is a *wave*. In order to appreciate the significance of this, we will have to take a slight detour into a more detailed discussion about what determines the speed of a wave.

12.4.1 THE SPEED OF WAVES

Because in relativity we are concerned with relative motion, we can ask how the motion, relative to the medium, of the source of a wave or the observer of a wave affects the measured speed of the wave. The answer is different, depending on whether we are talking about the motion of the source or the observer. By the way, when we are speaking of the *measured speed*, we are, of course, referring to the speed measured by the observer (of the wave).

Let us first consider the case when the observer is not moving with respect to the medium, but the source of the wave is. For instance, a trumpet player is running through the air and the observer is stationary in the air. In this case, the motion of the source (trumpet) has no effect on the measured speed of the wave. Why is that? Because, as we have just stated above, the speed of a wave is determined only by the properties of the medium. Thus, the motion of the source cannot affect the speed of the wave.

On the other hand, if the observer is moving, then there is motion with respect to the medium, so the observer must measure a different speed than someone who is at rest with respect to the medium. In this case, Equation 12.1 ($v = v' + u$) is relevant, where u would be the speed of the observer relative to the medium, v' would be the speed of the wave determined by the medium (as measured by someone at rest in the medium), and v would be the speed of the wave as measured by the moving observer.

Because of the unique relation of the speed of a wave and its medium, many times the speed of a wave is quoted without referencing it to some frame. In that case, what is meant is the speed of the wave relative to its own medium (i.e., the speed as measured by an observer at rest in the medium). So, if someone states that the speed of sound is 1,000 ft/s, what is meant is that, relative to the air, the speed of sound is 1,000 ft/s. Of course, if an observer is moving toward the source of the sound at, say, 100 ft/s, then that observer would measure a speed of 1,100 ft/s.

We should note here that the situation is different when measuring the speed of some projectile such as a ball. In this case, the measured speed is affected if either the thrower of the ball or the observer of the ball is moving. For example, consider a boy who can throw a ball at a speed of 20 ft/s when standing still. If the boy is running at a speed of 5 ft/s when he throws the ball forward, an observer will measure the speed as 25 ft/s. This is different from a wave, where the motion of the source has no effect on the measured speed of the wave. On the other hand, if the observer is moving toward the stationary boy at 5 ft/s, then the observer will measure the speed as 25 ft/s. For projectiles, it is the relative motion of the source (thrower) and observer that determines the measured speed. For a wave, the motion of the source has no effect. It is the motion of the observer relative to the medium that determines the measured speed of the wave.

AN EXAMPLE: MEASURING THE SPEED OF AN AIRPLANE

While this is not the way it is done, one can imagine using sound waves to measure the air speed of an airplane. Let us consider all airports having towers, which send out sound waves in all directions, with all planes having devices to measure the speed of the sound waves. Figure 12.2 shows a plane approaching a tower that is sending out a sound wave. The speed of sound in air is known to be 1,000 ft/s. The plane's detector will measure a speed of, say, 1,200 ft/s. The pilot then knows the plane is approaching the tower at a speed of 200 ft/s. Hopefully you see this intuitively, but let us check this using Equation 12.1. To ensure we do it correctly, first we want to define positive and negative directions. Let us take positive (+) as being to the right. Since, in the figure, the sound wave is moving to the right, $v = +1,000$, and the pilot also observes the sound wave moving to the right, so $v' = +1,200$. We want to find the speed of the plane, which is u. We write

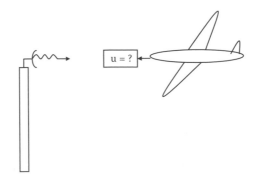

FIGURE 12.2 Plane approaching tower that is sending out sound wave. By measuring the speed of the wave, the pilot can determine the speed of the plane through the air.

$$v = v' + u \text{ or } u = v - v' = 1,000 - 1,200 = -200 \text{ ft/s}$$

In words from the ground observer's point of view, the plane is moving toward the left at 200 ft/s. If the plane were heading away from the tower at 200 ft/s, then the pilot would measure a speed of 800 ft/s, and from Equation 12.1, we would get

$$u = v - v' = 1,000 - 800 = 200 \text{ ft/s}$$

The pilot would say the plane is moving away (to the right) from the tower at a speed of 200 ft/s.

What this example is meant to show is that if the speed of a wave in the medium is known, and then that speed is measured by an observer, the speed of the observer through the medium can be measured. We will get back to this shortly.

12.4.2 THE ETHER

Let us now get back to Maxwell's discovery that light was a wave.

If light is a wave, then this should evoke a question in your mind. What's the question? Again, think about this before reading further.

The question you hopefully have asked yourself is: What is the medium for light? Now that we have the question, try to answer it.

There is a good chance that your answer was "the air." It is the answer many people would have. But a little thought should tell us that "the air" cannot be the correct answer. Why? Because we also know that light comes to us from the sun and stars outside our solar system and galaxy. Since the air is around us only very close to the earth, it cannot be the medium for light. There must be some other medium for light waves. This was exactly the thinking of physicists in 1865 after Maxwell showed that light was a wave. Even though they did not know what this medium was, they gave it the name *ether*. The ether is not very easy to detect. You cannot feel it or taste it or smell it. We know it should exist since light requires a medium to reach us from distant stars. In fact, while we cannot seem to detect any of its properties, we do know that it must be universal since we see light from the entire universe.

In the minds of the physicists in 1865, the fact that this medium was universal gave it a special significance, namely, that the ether was an *absolute reference frame* to which all speeds could be referenced. Wouldn't it be nice to know what the earth's true speed was? After all, the earth has a complicated motion. It moves around the sun at a speed of 18 miles/s; the sun moves around the center of the galaxy at 150 miles/s, carrying the earth with it; our galaxy moves with respect to our local group of galaxies; and our local group moves with respect to the distant galaxies. So, what is our true, universal, absolute speed? In other words, what is the speed of the earth with respect to the ether?

At first, this seems to be quite easy to answer. Remember our airplane example above. All we have to do is measure the speed of light just as the airplane pilot measured the speed of sound, since we are moving through the ether medium analogous to the airplane moving through the air medium. Also, it does not matter what the source of the light is. It could be light generated on earth or light from a star. This is because, as was stated above, the motion of the source of a wave has no effect on the measured speed of the wave.

As an example, let us say that the measured speed of light on earth is 186,100 miles/s. Remembering that Maxwell's speed for light (relative to the ether) is 186,000 miles/s, we can then calculate the speed of the earth relative to the ether as

$$u = v - v' = 186,100 - 186,000 = 100 \text{ miles/s}$$

In other words, this would be the absolute, universal speed of the earth. This sounds very straightforward and simple. In theory it is, but in practice there is a problem that has to do with experimental accuracy. In 1865 we did not have the technology we have today. Measuring speeds of the order of 100,000 miles/s was not so easy. Let us say one could achieve an accuracy of 1% (the error) in the measurement. In fact, this would have been much too accurate at that time, but let us use that number to make the point. An error of 1% means an uncertainty in the measured speed of 1,861 miles/s (0.01 × 186,100). We now quote the result with its uncertainty:

$$u = 100 \pm 1,861 \text{ miles/s}$$

What this indicates is that while we think we have measured the speed of the earth as being 100 miles/s, our experimental uncertainty in that number is 1,900 (we can round here). The error is about twenty times the measured number. In reality, for that time, the error would have probably been more than a hundred times the measurement. In other words, we really do not have any idea what the correct answer is since the uncertainty in that number is so much larger than the number itself. In fact, in general, this is one of the most inaccurate types of measurement to attempt, namely, one where the desired answer is the difference between two numbers that are much larger than that difference. The reason, as we saw above, is that the uncertainty is on the measured, large number, not on the desired small number. If the error was 1% of the desired value of 100, then the uncertainty would be 1 mile/s and we would have an accurate result. So, if an accurate measurement of the speed of the earth through the ether was to be achieved, a new technique was needed.

12.5   THE MICHELSON MORLEY EXPERIMENT

The new technique was devised by Albert Michelson, who was the first American scientist to be awarded the Nobel Prize. Before describing this ingenious experiment, let us make sure we are fully aware of an important aspect of the speed of light: how large a speed this is. We sometimes look at a number as just a number without realizing its full meaning. A speed of 186,000 miles/s is an enormous speed. Each second light covers a distance of 186,000 miles. Another way of making this point is to realize that in 1 billionth (10^{-9}) of a second light travels 1 ft. The fastest any human has traveled is the speed of an astronaut in orbit, which is 18,000 mph, or 5 miles/s. As a fraction of the speed of light, this is about $1/40,000 = 2.5 \times 10^{-3}\%$ of the speed of light. The point is that the speed of light is extremely large and not so easy to measure. Even today we can measure the speed of light fairly accurately, but it takes rather sophisticated equipment to do so.

Let us now consider Michelson's idea. Remember what was wanted was a measurement of the earth's speed through the ether. In the discussion above we tried to obtain this by indirectly measuring the speed of light (v) and then getting the desired answer (u) by subtracting the speed of light relative to the ether (v´), v´ (186,000 miles/s) being the speed Maxwell derived. Michelson's idea was to measure the desired quantity (u) directly. The method he devised to do this was ingenious. In fact, his technique is still used today where appropriate precision measurements are required. It is worthwhile going into the relevant details of the experiment. We will have to do some basic algebra, which you should approach as being part of a fascinating detective story. Hopefully when we are finished, you will have a greater appreciation as to the creativity involved in experimental science. In addition, the discussion will provide us another example of the methodology of physics.

12.5.1 AN ANALOGY: BOATS IN A RIVER

Michelson's basic idea was to set up a race between two light beams. In order to best understand this, we will first consider an analogous situation of a race between two boats in water. Instead of two light beams moving through the ether, we will have two boats moving through water. The water will be flowing in a river relative to the shore just as the ether is moving by the earth relative to us. After all, since the earth moves through the ether, from our point of view (i.e., relative to us), the ether flows by us. The boat-light race analogy is a good one, so the results we get for the boat race will be directly applicable to the light race of the real experiment.

The race will have two boats moving at right angles to each other. One will be moving up- and downstream in the direction the river is flowing, and the other will be moving cross stream, as illustrated in the Figure 12.3.

FIGURE 12.3 Two boats in a river. Each boat travels in still water with the speed v´, and the river moves relative to the shore with speed u. In the race both boats leave the start point at the same time, go to turnaround posts, each a distance d from the start point, and then return to the start point. The river flows in the direction of the right-hand arrow.

We want to see which boat gets back to the starting point first. Actually, we are really interested in the difference between the two times of arrival for the two boats since this is related to the speed of the river.

First, let us consider the case when the river is not moving ($u = 0$). If the start post and the distant post are a distance d apart, then using our old friendly equation $d = vT$, we see that each boat takes the same time, $t = d/v'$, to get to the far post and the same time to get back. So the total time to go from start to finish is

$$T_{total} = 2d/v'$$

In this case, when the river is not moving, both boats return to their starting point at the same instant, so the time difference is zero.

Now, what happens when the river is moving with speed u? We will derive the answer for the down/upstream trip and just state it for the cross-stream trip. For the down/upstream trip you might first guess that the answer is the same because the effect of the speed of the river just cancels out from going first downstream then upstream. Let us do it step by step and see.

For the downstream trip, the boat moves with a speed of $v' + u$ relative to the shore since it is moving in the same direction as the river. The time is:

$$T_{down} = d/(v' + u)$$

On the other hand, for the upstream trip the speed relative to the shore is now $v' - u$ since the boat is going against the flow of the river. This time is

$$T_{up} = d/(v' - u)$$

The total time is the sum of these two times:

$$T_{up/down} = T_{down} + T_{up} = d/(v' + u) + d/(v' - u) \qquad (12.2)$$

At this point we have to do some simple algebra of adding fractions, which we do by making the denominators the same by multiplying numerator and denominator of each term by the denominator of the other term. Remember, multiplying numerator and denomintor by the same amount does nothing to the value of the term; it is like multiplying by 1.0.

$$T_{total} = d(v' - u)/[(v' + u)(v' - u)] + d(v' + u)/[(v' + u)(v' - u)]$$

Note that we have now made the denominator of each term the same, so we can factor them out. In fact, that is the reason for doing the multiplication above.

$$T_{total} = d/[(v' + u)(v' - u)]\{(v' - u) + (v' + u)\}$$

The term $\{(v' - u) + (v' + u)\} = 2v'$. Also, note that $[(v' + u)(v'- u)] = v'^2 - u^2$. So, we can write:

$$T_{total} = 2dv'/[(v' + u)(v' - u)] = 2dv'/(v'^2 - u^2)$$

To make our answer look similar to the one we obtained when the river was not moving, we divide both numerator and denominator by v'^2 to get our final answer:

$$T_{up/down} = (2d/v')/(1 - u^2/v'^2) = (2d/v')/(1 - (u/v')^2) \qquad (12.3)$$

The numerator is identical to what we had for the time when the river was not moving, but now we have a denominator that is less than 1. That means that with the river moving, the up/downstream boat takes a longer time to get back than when the river is still. We can see why this is so by considering the special case when the river is flowing with the same speed as the boat moves in still water (i.e., $u = v'$). Then, when the boat turns around to go back upsteam, its speed relative to the shore is 0. In other words, it does not move at all, so it can never get back to its starting point.

For the cross-stream time, as was said above, we will just state the result:

$$T_{cross} = (2d/v')/\sqrt{(1 - (u/v')^2)} \qquad (12.4)$$

This looks almost like the expression for the up/downstream time, except there is a square root in the denominator. (For those of you who are curious where the square root comes from: it is due to the fact that for the cross-stream boat to get to its post, it cannot head directly toward it because of the flow of the river. The relevant speed is obtained by using the Pythagorean theorem [$v^2 = v'^2 - u^2$], and to get v, we have to take the square root.)

By noting a fact of arithmetic, we can see that the cross-stream time is always less than the up/downstream time. The denominator for T_{cross} is just the square root of that for $T_{up/down}$, and that expression is always less than 1.0. The arithmetic fact is that the square root of a number less than 1.0 is always greater than the number itself. (Try this for yourself with your calculator.) That means that the numerator in T_{cross} (which is the same as the numerator for $T_{up/down}$) is divided by a number larger than that for $T_{up/down}$. Hence, T_{cross} will always be less than $T_{up/down}$. In other words, the up/down boat always returns after the cross-stream boat.

Note that both times depend only on the values of d, v', and u, where d and v' are known numbers for the distance between posts and the speed of either boat in still water. So, if the times are measured, then the unknown speed of the river, u, can be calculated. Instead of doing the algebra to solve for u, let us put in some numbers to see how the times are affected. We will do this for two different values of u. For d choose 1 mile, and for v' choose 10 mph.

1. $u = 6$ mph:

$$T_{up/down} = 2d/v'/(1 - (u/v')^2) = (2*1/10)/(1 - (6/10)^2) = 0.2/(1 - .36) = 0.31 \text{ h}$$

$$T_{cross} = (2d/v')/\sqrt{1 - (u/v')^2} = 0.2/\sqrt{1 - .36} = 0.2/\sqrt{.64} = 0.2/0.8 = 0.25 \text{ h}$$

The time difference is $.31 - .25 = 0.06$ h $= 3.6$ min.

2. u = 8 mph:

$$T_{up/down} = 0.2/(1 - (8/10)^2) = 0.2/.36 = 0.56 \text{ h}$$

$$T_{cross} = 0.2/\sqrt{(1 - .64)} = 0.2/\sqrt{.36} = 0.2/0.6 = 0.33 \text{ h}$$

Now the time difference is $0.56 - 0.33 = 0.23$ h $= 13.5$ min.

We see the time difference is quite sensitive to the speed of the river.

At this point, you may be asking why use the time difference? After all, either the up/down or cross-stream time itself depends on u. The reason has to do with the actual Michelson experiment. The technique that he came up with measures only the time difference.

12.5.2 THE REAL EXPERIMENT

As was stated above, in the actual experiment the race was between two light beams. Here the light is analogous to the boats, with the speed now being $v' = 186,000$ miles/s, the speed of light with respect to the ether. The speed to be measured, u, was the speed of the earth through the ether, or from our point of view here on earth, the speed of the ether wind "blowing" by the earth. The diagram of the experimental apparatus is shown in Figure 12.4. Instead of boats there are two light beams, and instead of turn-around posts there are mirrors. The apparatus is known as an *interferometer* because it is based on the wave property of interference as discussed in the last chapter. The half-transparent mirror allows half of the light to go through and and half to be reflected, creating two light waves that can interfere with each other when they recombine. Here we now want to get an idea of how sensitive it is.

What time difference did Michelson expect? Remember, he wanted to be able to measure the speed of the earth through the ether. Since he did not know the answer before the experiment, he hoped he could at least measure a speed comparable to the earth's speed around the sun, which is 18 miles/s. Let us see what time difference this corresponds to. We will do exactly what we did before with the boats in the river. In order to get our answer, it will take the use of an approximation and a bit more algebra because we will be dealing with a number like 186,000 miles/s instead of 10 mph. But it is straightforward and should be easy to follow. In fact, you will get more out of this if you do not just read it, but follow along with pencil and paper.

We start by using Equations 12.3 and 12.4 to get the expression for the time difference, but using the symbol c for the speed of light instead of v' for the speed of the boats.

$$\Delta T = (2d/c)/(1 - (u/c)^2) - (2d/c)/\sqrt{(1 - (u/c)^2)}$$

$$= (2d/c)\{1/(1 - (u/c)^2) - 1/\sqrt{(1 - (u/c)^2)}\} \qquad (12.5)$$

At this point, let us look at the ratio u/c. If we use $u = 18$ miles/s and $c = 186,000$ miles/s, the ratio is

$$u/c = 18/186,000 \approx 1/10,000 = 10^{-4}$$

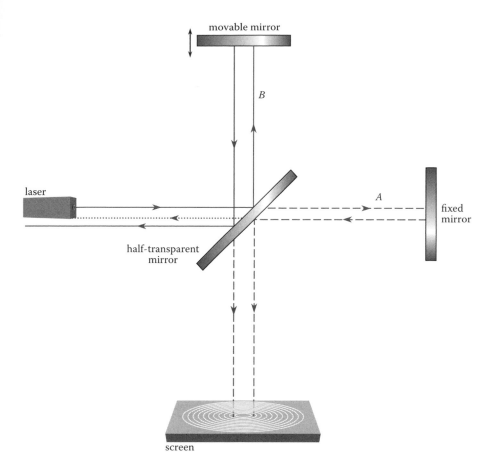

FIGURE 12.4 The Michelson interferometer. As the name implies, Michelson used the interference of waves, which was discussed in Chapter 11. Light from a single source is split by the half-transparent mirror causing it to go in two perpendicular directions. This ensures that the two beams start off at exactly the same time. The mirrors reflect the light back to their starting point, allowing the two light beams to combine and interfere. A time difference of a few percent of the period of the light used could be measured. The movable mirror was used to align and calibrate the interferometer. (From Encyclopedia_Britannica;www.britannica. com/cdchecked/topic_art/340440)

where we have used $\approx$ to indicate 10^{-4} is approximate but certainly good enough for our purposes. The number we need is $(u/c)^2$, which is 10^{-8}. This is a very small number compared to 1.0. So, even using your calculator, subtracting 10^{-8} from 1.0 will still give the answer 1.0. Then Equation 12.5 would give zero for the time difference. While the time difference is indeed small, as we will see below, it is not exactly zero. Here is where we will introduce a very useful approximation to allow us to find the correct time difference.

For a small number, call it x, we can write the following:

$$\sqrt{(1-x)} \approx 1 - \tfrac{1}{2}(x) \qquad (12.6)$$

$$1/(1-x) \approx 1 + x \qquad (12.7)$$

You should be able to verify these for yourself with your calculator for small values of x. For instance, if $x = 0.1$, then $\sqrt{.9} = 0.949$ and the approximation gives 0.95, which is a difference of 0.1%. Also, $1/.9 = 1.111$ while the approximation is 1.100, which is a difference of 1%. The smaller the value of x, the more accurate the approximation. Try it for yourself with $x = 0.01$. For $x = 10^{-8}$, the approximation is very accurate. We now write Equation 12.5 as

$$\Delta T = (2d/c)\{1 + (u/c)^2 - (1 + \tfrac{1}{2}(u/c)^2)\}$$

$$= (2d/c)\{1/2(u/c)^2\} = d/c(u/c)^2 \qquad (12.8)$$

This is a very simple expression where all we have to know is d, the length of the interferometer arm. Michelson chose this to be 1.2 m. Thus, we have

$$\Delta T = (1.2/(3 \times 10^8))(10^{-8}) = 0.4 \times 10^{-16} \text{ s}$$

($c = 3 \times 10^8$ m/s was used above to get the correct units). Given that the period of green light is 1.7×10^{-15} Hz, this time is just about 2.5% of the period of the light ($.4 \times 10^{-16}/1.7 \times 10^{-15} = 0.025$), which was about what Michelson thought he could measure.

There is one other detail that must be mentioned. Because of all the different motions the earth is undergoing, Michelson had no idea of the true direction of motion through the ether. In the words of our boat analogy, he did not know what was up- or downstream or cross stream. He fully realized this, so he mounted the interferometer in a big vat of mercury so he could easily rotate the entire apparatus. In fact, what he expected was that he would observe a time difference between different orientations of the interferometer.

Michelson performed the experiment in 1881. The result was that he could not detect any measurable time difference for any orientation. This was not what was expected. So, what do we do now? In some respect, this is similar to the predicament that we had when discussing beta decay in Chapter 10, where the measured energy distribution for the electrons did not agree with the expected distribution. In that case, we saw that the first explanation was that the experiment was just wrong. For the Michelson experiment the situation is somewhat different since Michelson had calibrated his intererometer introducing known time delays that he could then measure. So, he knew the apparatus was working correctly. But maybe it was not sensitive enough to measure the effect. After all, it was designed to be able to just barely measure a time difference corresponding to the speed of the earth around the sun. What if the true speed was less than that?

The problem Michelson faced was how to make his interferometer more sensitive. If we look at Equation 12.8, we see that the only quantity that Michelson could change was d, the length of the interferometer arms; u was what he was trying to measure, and the value of c, the speed of light, is fixed by nature. The larger the value of d, the larger the time difference would be. To make sure the interferometer would be sensitive enough, it would be necessary to make a large change in the size of the arms. It turns out that was not so easy to do. In order to accomplish this, he asked his friend Edward Morley for help. They increased the distance d not by making longer arms, but by increasing the distance the light traveled by using multiple reflections. It took 6 years to perfect the new interferometer. The distance, d, was increased to 11 m, almost ten times the original distance. The now famous Michelson Morley experiment was performed in 1887 with a sensitivty almost ten times greater than the original.

The results of that experiment were the same as the first. Namely, no observable time difference could be observed for any orientation of the interferometer. In other words, it seemed that the two light beams always came back at the same time. How could this be?

One possible explanation was that the earth did not move through the ether, i.e., it was at rest! After all, the time difference was a measure of the earth's speed through the ether. If there was no time difference, then doesn't that tell us that the earth is not moving? This would certainly make the earth a rather special place in the cosmos.

> But there is a big problem with this explanation. Do you see it? Try to think about this before reading on. As a hint, think about what you do know about the earth's motion.

Figure 12.5 below should help us to understand the problem. It shows the earth going *around* the sun. We also know the sun goes around the center of our galaxy, the galaxy moves with respect to the local group of galaxies, and the local group moves with respect to the distant galaxies. The arrow at the sun represents the net velocity of the sun relative to the ether, and the arrows at the top and bottom positions of the earth (6 months apart) represent the velocity of the earth relative to the sun (18 miles/s).

If the earth were not moving with respect to the ether, it would have to be for the case when the earth is at the top of the figure. Also, the sun's speed relative to the ether would have to be 18 miles/s. In that case, since the two velocities would be equal and opposite, they would cancel making the earth's speed relative to the ether identically zero. But as the figure shows, 6 months later the earth would be moving in the same direction as the sun and then the speed of the earth relative to the ether would be 36 miles/s—a speed that the Michelson Morley apparatus could easily measure.

The experiment was repeated at differing times throughout the year with always the same effect. The two light beams returned at the same time.

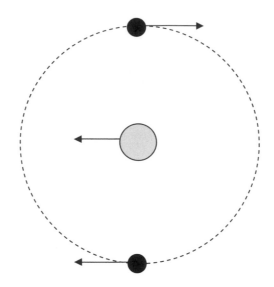

FIGURE 12.5 Diagram depicting the earth's motion around the sun and the sun's speed with respect to the ether.

12.5.3 THE LORENTZ CONTRACTION

This was certainly a great surprise. Physicists at the time could not understand this strange result in terms of any known physics. When such a situation arises, new ideas, not based on already known physics principles, are introduced. This can lead to new theories, which result in an entire revolution in our thinking or a specific and limited idea to explain a particular experimental result. These ideas are usually known as ad hoc. In Chapter 9, we discussed Bohr's ad hoc assumptions about the atom. Ad hoc ideas or assumptions have been quite useful throughout the history of physics. Eventually some have been shown to be correct, while others have not. We will discuss several more in this book.

After the Michelson Morley experiment, H. Lorentz made an ad hoc sugges-tion to explain the strange results. His idea was that there was an unrealized effect on lengths of objects as they moved through the ether. This effect caused the lengths in the direction of motion through the ether to shorten, or contract. Lengths in the direction perpendicular to the motion through the ether would not be affected. There was absolutely no theoretical basis for this suggestion. But as we shall see below, it did explain the null result of the Michelson Morley experiment.

One way of trying to explain why this should happen is by the following argument: The ether is an electromagnetic medium—after all, it is the medium for light. When objects move through the ether, the electric force only in the direction of motion is increased. Since atoms are held together by the electric force, the atoms along the line of motion through the ether are pulled closer together. This would cause the length of an object in that direction to contract. It is still ad hoc that the electric force increases, but it is a physical way of explaining the contraction. This is just an example of how one could think about what might be happening to cause the effect.

How does Lorentz's sugestion explain the null result of the Michelson Morley experiment? First, you have to know that he made a very specific and quantatative suggestion as to the contraction. If we write d_c as the length with contraction, then the Lorentz contraction equation is

$$d_c = d\sqrt{(1 - (u/c)^2)} \tag{12.9}$$

Since the square root is less than 1.0, d_c is less than d; i.e., the length is contracted. If we put this expression into Equation 12.3 (but using c instead of v') for the up/down time (the time along the direction of motion through the ether), it becomes:

$$T_{up/down} = (2d_c/c)/(1 - (u/c)^2) = 2d\sqrt{(1 - (u/c)^2)}/(1 - (u/c)^2) = (2d/c)/\sqrt{(1-(u/c)^2)}$$

This is now identical to Equation 12.4 for the cross-stream time. Thus, the two times are equal and the light beams will always return at the same time independent of the speed of the earth through the ether. It was certainly a strange but ingenious idea.

12.5.4 ANOTHER CRAZY IDEA

While we are discussing weird ideas, let us consider one other that at first looks even stranger than the Lorentz contraction. There is another way, besides Lorentz's idea, to always have the two beams come back at the same time. Going back to our discussion of the boats in the river, we saw that the speed of the boats is affected by how fast the river is moving. Going downstream, the speed is v' + u, and upstream it is v' − u. For the actual Michelson Morley experiment, it was the speed of light that is relevant. Thus, it was supposed that the speed of light when the light was moving with the ether would be c + u, and c − u when it was moving opposite to the ether. This is exactly what Galilean relativity predicts.

But what if the speed of light is always c? In other words, the motion of the earth through the ether did not affect light speed. If this were the case, then the two light beams would always have to come back at the same time since the path they traveled and their speed would be identical. This would then explain the null result of the Michelson Morley experiment. But, let us look at some of the consequences of this idea.

Consider Figure 12.6, where two rocket ships are approaching each other with relative speed u. There is a source of light on one of the ships and an observer on the other. According to Galilean relativity, the observer measures the speed of light as being

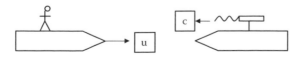

FIGURE 12.6   Two rocket ships approaching each other with relative speed u. A source of light with v' = c on one ship is directed at an observer on the other.

$$v = v' + u = c + u$$

But the hypothesis we are considering says the speed of light is independent of the motion of the source or observer. This means $v = c$ also. This makes no sense since if we were to put it into our equation above, we would have:

$$c = c + u$$

The only way this equation could be true is if u were zero. But u is the relative speed of two rocket ships and could be anything. This is ludicrous. To make the point even more dramatic, let the two ships approach each other at the speed of light, i.e., $u = c$. Then we can write:

$$c = c + c \text{ or } c = 2c \text{ or } 1 = 2$$

which is certainly nonsense!

So, why are we discussing this crazy idea that the speed of light is the same for all observers if it leads to such a nonsensical result? Because we now know it is correct! If this is so, then it must mean that our Galilean relationship, $v = v' + u$, cannot be correct! The rest of this chapter will be devoted to understanding this.

12.6 ASSUMPTIONS WE TAKE FOR GRANTED

Our discussion above has led us to be concerend about how two observers moving with respect to each other measure the speed of light. In order to try to understand this, let us set up the most simple possible experiment to measure the speed of light as depicted in Figure 12.7.

Since speed is a measure of *distance* traveled in a given *time*, we need devices to measure distance and time, in other words, a ruler and a clock. The figure below indicates the experimental setup. Each observer is to measure the speed of light using accurate rulers and stopwatches. Observer A, at the top, is at rest relative to us, while observer B is moving at half the speed of light. The dashed vertical line represents the starting line where A is positioned and both observer B and the light

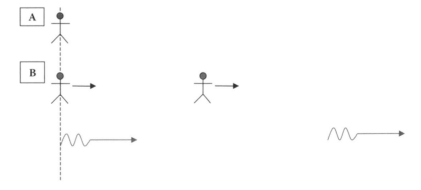

FIGURE 12.7   The speed of light measured by observer A, who is at rest with respect to us, and observer B, who is moving at half the speed of light with respect to us. Both observers measure the distance the light beam is from them after the time of 1.0 s. The speed of light is obtained from the equation $v = d/t = d/1.0$.

beam cross at time $= 0$. That is the time when both A and B start their stopwatches. Afer 1.0 s has elapsed on both their watches, they will measure the distance the light beam is from them. We want to calculate the speed each finds using his or her ruler to measure his or her distance from the light at the time of 1.0 s as measured on his or her clock. The speed is then obtained from our old friend $v = d/T$.

1. Speed as measured by A:

$v_A =$ (distance from light to A)/time $= 186,000/1.0 = 186,000$ miles/s

2. Speed as measured by B:

$v_B =$ (distance from light to B)/time $= 93,000/1.0 = 93,000$ miles/s

It appears that A and B have measured different values for the speed of light, just as we might expect. But this contradicts our statement above that all observers, independent of their relative motion, always measure the speed of light as being the same. What is the matter with our simplest of experiments described above?

The problem is that in doing the analysis above, we have made three different assumptions that are so ingrained in our thinking that we do not even realize that any assumptions are being made. If any one of these assumptions is wrong, then the results of that analysis are wrong.

Try to identify those assumptions before reading further.

Assumption 1: Both A and B agree that they started and stopped their stopwatches *at the same time*. In other words, they could do this *simultaneously*. If they could not, then they would not agree that the elapsed time would be 1.0 s.

Assumption 2: Both the A and B clocks keep the same time. In other words, one did not run slow or fast with respect to the other—they remained calibrated independent of their relative motion. If this were not true, then 1.0 s would be different for each observer. Note that this is different than the first assumption even though both have to do with time. Here the clocks would be seen as running at different rates, while for assumption 1 the disagreement has to do with when the clocks were started and stopped.

Assumption 3: Both the A and B rulers remained the same length. If the distance is measured incorrectly, then the speed will be measured incorrectly.

As was said above, if any one of these assumptions is wrong, then our analysis would not be correct. The truth is all three of the assumptions are wrong. Understanding why this is so requires us to understand Einstein's theory of relativity. At first this may sound rather daunting, especially for someone who is not a scientist. Hopefully you will see that the ideas of relativity are quite logical and understandable, even though some of the predictions of relativity seem rather far out. By the end of our discussion, you should have a good appreciaton and understanding of the ideas of relativity.

12.7   THE POSTULATES OF SPECIAL RELATIVITY

Einstein's theory of relativity is based on two postulates:

All inertial reference frames are equivalent.

The speed of light is the same for all observers independent of their relative motion.

The first postulate should look very familiar. It is just the postulate of Galilean relativity. The second postulate is just the "crazy" idea we discussed in Section 12.5.4. We will see where these lead us shortly.

But first let us look at an immediate consequence of the second postulate: there is no ether! Remember, the pupose of the Michelson Morley experiment was to detect the earth's motion through the ether. But given postulate 2, there is no way to do this. Combining this with the fact that the ether was not detectable through any of our senses means that there is no way of detecting the ether. In science, if there is no way of detecting something, directly or indirectly, then it makes no sense to believe that it exists at all.

At this point, you may be asking yourself the following question: If there is no ether, what is the medium for light? After all, light is a wave and a wave requires a medium to be disturbed. Well, in the case of light waves, there is a medium to be disturbed. But it is unlike all other waves in that it is not a material medium. In other words, it is not a medium made up of atoms. For light, what is being disturbed are electric and magnetic fields. Light is an electromagnetic wave that is due to changing electric and magnetic fields. In Chapter 6, we first noted that in our discussion of the change in the electric field when charged particles were accelerated. In the table below, properties of three different electromagnetic waves are shown. While they are all electromagnetic waves, their properties and how they interact with matter are quite different. It is the frequency, or wavelength, that distinguishes them. The table shows the relevant information of three different electromagnetic waves. The only difference is the frequency or wavelength, yet we know they have very different properties.

	Radio Waves	Visible Light	Gamma Rays
λ	300 m	$7 \times 10^{-7} - 4 \times 10^{-7}$ m	$<10^{-12}$ m
f	10^6 Hz	$4 \times 10^{14} - 7 \times 10^{14}$ Hz	$>10^{20}$ Hz
T	10^{-6} s	$2.5 \times 10^{-15} - 1.4 \times 10^{-15}$ s	$<10^{-20}$ s
Energy	10^{-8} eV	$1.7 - 2.9$ eV	$>10^6$ eV

12.7.1   SOME INTERESTING FACTS ABOUT EINSTEIN AND THE BIRTH OF RELATIVITY

Einstein published his paper on relativity in the year 1905. At that time, he was a patent examiner in the patent office in Bern, Switzerland. He claimed the job gave him time to think about the nature of nature. Even though he was completley unknown, in that single year he published five papers. Many feel any one of them was worth a Nobel Prize. In fact, Einstein never received a Nobel Prize for his relativity paper. He did receive one in 1922 for another paper he published in 1905. It was for what is known as the photoelectric effect, which explains, for instance, how the exposure meter works in a camera. It was a much more controversial paper than the one on relativity. It hypothesized what turned out to be one of the founding ideas in quantum theory. We will discuss it in the next chapter.

The story we have told so far might lead you to believe that it was the null result of the Michelson Morley experiment that led Einstein to his ideas about relativity. It turns out that is not the case. His paper on relativity does not reference that experiment. It is not clear that Einstein knew of the experiment at all. If he did, he certainly did not pay much attention to it. Einstein has, in fact, talked about his thinking that led him to his ideas. According to him, when he was about 16 years old he started questioning what it would be like to be able to catch up to a light beam.

The word *relativity* does not appear in the title of the paper. The title was "On the Electrodynamics of Moving Bodies." Interestingly, that now famous and important paper begins with a question: "Why is there in Maxwell's theory one equation for finding the electromotive force generated in a moving conductor when it goes past a stationary magnet, and another equation when the conductor is stationary and the magnet is moving? After all, it is only the relative motion between magnet and conductor that counts." We have not talked about electromotive force. It is due to Faraday's law. When a conductor moves in a magnetic field, it experiences a changing field that induces an electric field, which causes the electrons in the conductor to move. The moving electrons then experience a Lorentz force due to their motion in the magnetic field that is present.

Einstein could have also asked a similar question having to do with Ampere's law, which, remember, states that moving electric charges produce a magnetic field. But if an observer is moving along with the charges, then that observer is at rest with respect to those charges. Then from that observer's point of view, there are no moving charges, and hence there should be no magnetic field. Well, is there a magnetic field or not? It is questions like these that led Einstein to create one of the great revolutions in our understanding of the physical universe.

12.8 CONSEQUENCES OF THE POSTULATES OF RELATIVITY

Now that we have stated the two postulates of relativity, we will be able to understand why the three assumptions discussed above are incorrect. Everything we will discuss follows from these postulates and some careful thinking.

12.8.1 THE RELATIVTIY OF SIMULTANEITY

What do we mean when we say two events are simultaneous? The most obvious answer is that the two events happen at the same time. But we have to be careful with this "obvious" answer. When an observer claims an event has occurred, many times he or she is not at the location of the actual event. What is being observed is the signal caused by the event (sound, light, etc.) that reaches the observer. That signal takes time to travel from the event to the observer. That time depends on the location of the event and the speed of the signal. To see the consequences of this, let us consider Figure 12.8. Two observers are positioned with respect to two loudspeakers, as shown. The switch centered between the loudspeakers is closed, causing a pulse of sound to be emitted from each speaker.

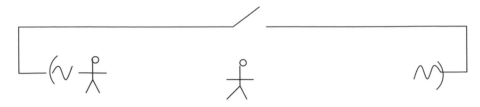

FIGURE 12.8 Two loudspeakers will emit sound waves when the switch midway between the two speakers is closed. The two observers are located such that one is close to the left-hand speaker while the other is midway between the two speakers.

It is clear that the observer close to the left-hand speaker will hear the sound from that speaker before the sound from the far, right-hand speaker, while the observer midway between the speakers will hear both sounds at the same time. So, even in the case when the two observers are at rest with respect to each other, they do not agree that the two events (their reception of the sound) occurred at the same time. But in this case, there is really no ambiguity. From the setup of the equipment, we know that the two sounds did indeed leave the speakers at the same time. It is just that the left-hand observer is closer to one speker and farther from the other. We can unambiguously define events to be simultaneous if an observer midway between the two events detects them at the same time. So, in the case of determining whether events are simultaneous, only observers midway between the events are valid judges.

Since relativity has to do with observers in motion, now let us consider what happens when we have the two observers moving with respect to each other. Figure 12.9 illustrates this case. Two observers in their own rocket ships have been wandering through the galaxy looking for simultaneous events. Both rocket ships are inertial reference frames. They are passing by each other just as two meteors hit the ends of the ships. The fact that the ends of the ships were hit can be verified at any time later. Observers A and B are exactly in the middle of their respective rockets. Thus, they are midway between the two events and are both valid judges as to whether the events (the two meteors striking the ends of the rockets) are simultaneous, or not.

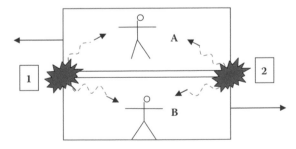

FIGURE 12.9 Two inertial rocket ships pass by each other with relative speed u. As they are abreast, two meteors hit the ends of each ship (events 1 and 2) as indicated. Observers A and B are in the middle of their respective ships. The dashed lines represent the light caused by the collision, which is directed toward the appropriate observer.

In order to clearly understand our final conclusion, we want one of the observers to claim the two events are simultaneous. We will see it does not matter which one makes that claim, so let us say it is observer A. In other words, A says that the two light signals from the events reached him at the same time. He joyfully radios to B that he has finally observed two simultaneous events, and he is a good observer of simultaneity because he is midway between the two events.

Now let us analyze what observer B claims. She says that A is being fooled by the fact that he is moving and only thinks the two events were simultaneous, but in actuality, they were not. Why does B say this? Because, from her point of view, A is moving to the left, and hence into the light coming from event 1 and away from the light coming from event 2. According to B, event 2 happened before event 1, but the two light signals *reached* A at the same time due to his motion. In other words, B agrees with A that the signals reached him at the same time, but they did not *leave* the events at the same time.

Observer A now retorts back. He says to B: "You are mistaken. I am at rest and you are moving by me to the right. The two events are simultaneous, but I understand why you claim they are not. Since your motion is to the right, you are moving toward the light from event 2 and away from the light from event 1. I agree that you did receive the signal from event 2 before that from event 1, but that is because of your motion. The two signals did leave at the same time."

Who is right? Hopefully you see that both are right. They are both valid observers of simultaneous events since they are both midway between the events. Each claims that he or she is at rest and the other one is moving. Since they are both in inertial reference frames, one frame is no better or worse than the other (the first postulate of relativity). There is no way to choose between them.

From this reasoning, we are forced to the conclusion that there is no *absolute* way of determing whether two events are simultaneous. If two observers are moving with respect to each other, then they cannot agree that events are simultaneous if the two events are separated in space, as in our example. The only way there will be agreement is if the two events occur in the same place. Then, if an observer claims the events are simultaneous, all observers will agree.

> To check that you have a good understanding of these ideas, analyze what A would say if B claimed both events were simultaneous. You may want to discuss this in class.

Going back to our three assumptions, we now see that the first one about agreement on simultaneity is not correct. We will see that the fact that observers will not agree on simultaneity is key to understanding several supposed paradoxes in relativity.

12.8.2 TIME DILATION

Time can be rather hard to define, for instance, a nonspatial continuum in which events occur or, more simply, the passage of events. From an operational point of view, it can be defined as that which a clock measures. Then, of course, we have to define *clock*. That is easier. A clock is something that repeats itself due to a regularly occurring event. In order to understand the relativity of time, we want to be able to see what happens to a clock as it moves by us. There are all sorts of clocks: digital (probably your wristwatch), analog (a wall clock), or atomic clocks. While these can all be quite accurate (especially the atomic clock), they are also complicated. It would be very hard to analyze the effect that motion has on them. If we could find an accurate and simple clock to analyze, then whatever we find for that clock must be true for all other clocks. Assuming accuracy, all clocks must remain calibrated and do the same thing as long as they are in the same reference frame.

12.8.2.1 The Light Clock

Figure 12.10 shows our accurate and yet simple clock. It is known as a light clock because the regularly occurring event is a light pulse traveling between two fixed mirrors.

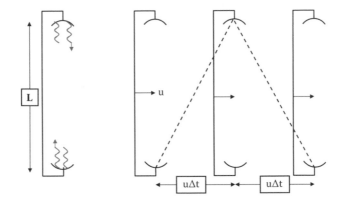

FIGURE 12.10 The light clock. In the first frame, the clock is at rest with a light pulse starting at the bottom mirror, reflecting from the top mirror back to the bottom mirror. The next three frames show three positions of the same clock if it were moving with respect to us with a relative speed of u. The last two positions indicate that the clock has moved a distance $u\Delta t$ each time the light has reached the appropriate mirror. The dashed lines represent the path of the light as it would appear to an observer who sees the clock move by.

The light clock has a well-defined period (time to get from the bottom mirror and back). It is obtained by using our old friend $d = vT$, or solving for the time, $T = d/v$. Using $d = 2L$ and $v = c$, we get for the time unit of our clock:

$$T_0 = 2L/c \tag{12.10}$$

This is the time as measured by an observer who is at rest with respect to the clock.

For an observer who sees the clock move by, the period is $2\Delta t$, where $\Delta t = H/c$, H being the hypotenuse of the triangle, one of whose sides is L. We can write

$$T = 2H/c \tag{12.11}$$

It is important to note that we have still used c for the speed of light even though the clock is moving by us. This is just applying the second postulate of relativity that states that *all* observers measure the speed of light as being the same. If we had applied Galilean relativity, we would have used a larger value for the speed of light, and T and T_0 would be equal. Since the speed is the same but H is greater than L, T is greater than T_0. In other words, the observer who sees the same clock move by measures a longer time than the observer who is at rest with respect to the same clock.

By looking at Figure 12.10 and doing a bit of geometry, we can find the correct relation between T and T_0. If you feel you do not want to go through the calculations below, you can just skip to Equation 12.12, which gives the final result. It is known as the time dilation equation, and we will be using that. But if you would like to see how we arrive at the equation, read on.

From Figure 12.10, we can get an expression for H that is the hypotenuse of the triangle whose sides are L and $u\Delta t$. Thus, we can write

$$H^2 = L^2 + (u\Delta t)^2$$

From Equation 12.11, $\Delta t = T/2 = H/c$, which allows us to write

$$(H/c)^2 = (\Delta t)^2 = \{L^2 + (u\Delta t)^2\}/c^2$$

Gathering the Δt's gives

$$(\Delta t)^2\{1 - (u/c)^2\} = (L/c)^2$$

Solving for $(\Delta t)^2$ and taking the square root gives

$$\Delta t = (L/c)/\sqrt{\{1 - (u/c)^2\}}$$

But $\Delta t = T/2$ and $L/c = T_0/2$, so we can write our final result:

$$T = T_0/\sqrt{\{1 - (u/c)^2\}} \qquad (12.12)$$

This is the time dilation equation, which tells by how much the time measured by an observer who sees a clock move differs by the time on the same clock but observed by someone who is at rest with respect to that clock.

There are several things to note about this equation:

1. The denominator always has a value less than 1.0. Hence, T is always greater than T_0, as we said above.
2. The quantity that determines the difference between T and T_0 is the ratio (u/c).
3. For normal speeds, the value of (u/c) is a number much less than 1.0, so $(u/c)^2$ is even a smaller number. For such a small number, the difference between 1 and $\sqrt{\{1 - (u/c)^2\}}$ is negligible, so the square root factor is also equal to 1. Thus, at these small speeds, T and T_0 are essentially equal, so the time dilation effect is not detectable. As an example, the fastest any person has traveled is 18,000 mph, which is 5 miles/s—the speed of an astronaut in

orbit. For this speed, $(u/c)^2 = (5/186,000)^2 = 7.2 \times 10^{-10}$. This explains why we do not experience time dilation in our ordinary lives.

4. When speeds get very close to c so that the ratio (u/c) gets close to 1.0, then the difference between T and T_0 can be substantial. For instance, at a speed of 150,000 miles/s, which is close to 80%, the speed of light (u/c = 0.8)

$$\sqrt{\{1 - (u/c)^2\}} = \sqrt{\{1 - .8^2\}} = \sqrt{\{1 - .64\}} = \sqrt{.36} = 0.6$$

or

$$1/\sqrt{\{1 - (u/c)^2\}} = 1/0.6 = 1.67$$

or

$$T = 1.67T_0$$

Let us make sure we understand the meaning of this result. If an observer on rocket ship A measures a time of 1 min for some event that occurs on his rocket, then observer B, who is on a rocket with a relative speed of 80% the speed of light, will measure a time of 1.67 min for that same event. Of course, if observer B measures a time of 1 min for an event in her rocket, then observer A will measure that event as taking a time of 1.67 min.

12.8.2.2 Useful Definitions

There are several phrases and quantities that continually appear in relativity. It is very convenient to define them here so we can use them from now on:

1. **Proper time:** The time as measured by the observer who is at rest with respect to the event. As we have seen, in relativity, there are usually two observers who are moving with respect to each other. An event, which takes a certain amount of time, occurs in one reference frame. The observer in that reference frame measures the proper time. We will call this person the *proper observer*. The observer who is moving relative to the proper observer will measure a longer time for the same event. The time measured by that observer we will call the *dilated time*. (Note that the word *proper* here does not imply any sort of value judgment. The time measured by the other observer is not an *improper* time. They are both valid observers. According to what we have shown above, they just measure different times.)

2. **β:** We noted above that the important quantity that time dilation depends upon is the ratio of u/c. This ratio appears so much that we give it a special symbol:

$$\beta \equiv u/c \tag{12.13}$$

The $\equiv$ means this is a defining equation.

3. **γ:** Looking at Equation 12.12, we see the factor that actually determines the amount of time dilation is $1/\sqrt{\{1 - (u/c)^2\}} = 1/\sqrt{\{1 - \beta^2\}}$. We define:

$$\gamma \equiv 1/\sqrt{\{1 - \beta^2\}} \qquad (12.14)$$

It of course depends only on β, but when β is small, γ is essentially equal to 1.0 and does not change for a large range of values for β. On the other hand, when β gets close to 1.0, γ can change very rapidly as β varies. γ appears in many relativistic relations and does determine the size of the effect. For that reason, it is sometimes known as the relativistic factor. Figure 12.11 is a graph illustrating how γ varies as a function of β. It would be very instructive for you to calculate a few of these values for yourself. For instance, the highest value of γ on the graph is for a value of γ equal to 0.99.

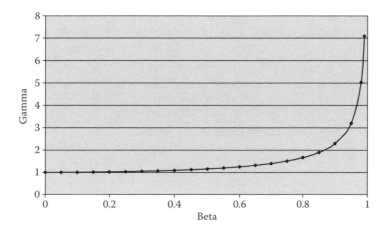

FIGURE 12.11   Graph of γ (gamma) versus β (beta).

Given our new definitions, we can rewrite Equation 12.12 as

$$T = \gamma T_0 \qquad (12.15)$$

With this new relation, a graph of T as a function of β would have the same shape as that shown in Figure 12.11.

Should we believe this? You may be wondering why we should believe any of this. Einstein was a genius, but that in itself is not sufficient reason to believe anything in science. What is required is experimental verification. But, you may say, if for the fastest speed any human has achieved, the predicted effect is not measurable, how can time dilation be verified?

While humans have not been able to achieve speeds close to the speed of light because of our relatively large mass, there are nuclear and subnuclear particles that do attain speeds very close to c—either by natural means or in man-made particle accelerators. One example is a fundamental particle known as the muon (symbolized by the Greek letter μ). The muon is essentially a heavy electron with a mass about 200 times that of the electron (106 MeV compared to 0.51 MeV for the electron). It decays into an electron and two neutrinos with a lifetime of 2×10^{-6} s. While this appears to be a very small time, it is actually quite easy to measure with high accuracy. Several muons per second are passing through you as you read this. They are the main constituents of *cosmic rays* (particles that impinge on earth that come from the sun or outside our solar system) at sea level on earth. Because of their very small mass, muons can achieve speeds very close to c, thus having values of γ that can be much greater than 1.0. Using particles such as these, time dilation has been very well verified.

12.8.3 LENGTH CONTRACTION

We have now seen that the first two of our assumptions, as to whether two observers will agree about determining time, are incorrect. We are now ready to consider our assumption about length. To do this, we are going to use the muon that was introduced above. The muons that reach us at sea level are primarily produced by interactions by particles from outer space, such as protons that interact in the upper atmosphere. Thus, we can say they are born high above us and at some point afterwards die by decaying into an electron and neutrinos, as described above. Let us consider one such muon that is born at the top of Mt. Everest, which is 5.5 miles high. This is illustrated in Figure 12.12.

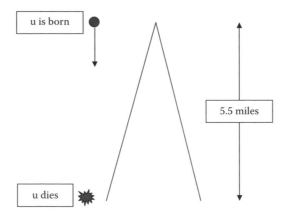

FIGURE 12.12 The muon and Mt. Everest. A muon is born at the top of Mt. Everest by the interaction of a cosmic ray with the air. It travels at a speed very close to the speed of light and dies (decays) at the bottom of Mt. Everest.

We will consider what is happening first from the point of view of an observer at rest on earth, and then from the point of view of an observer moving with the muon.

The vast majority of muons are highly relativistic, which means their speed is very close to the speed of light as observed on earth. Let us calculate how long it takes the muon to travel from the top to the bottom of Mt. Everest—in other words, how long it lived. We do this by again using our old friend d = vT. Solving for T and using c for v, we get T = d/c = 5.5/186,000 = 5.5 (1.86×10^5) = 3×10^{-5} s = 30×10^{-6} s.

Do you have a question? Given the lifetime for the muon we just calculated above, you should. If you see the question, do you see the answer? Again, try to think about this before reading further.

When the information about the muon was introduced above, it was stated that its lifetime is 2×10^{-6} s. But the lifetime we have just calculated for our muon going by Mt. Everest is 30×10^{-6}, or fifteen times longer. Hopefully you see that this is just an example of time dilation due to the fact that we observe the muon moving relative to us with a speed very close to c. In fact, the factor of 15 is just γ. The time of 2×10^{-6} is the proper lifetime. When lifetimes are quoted for decaying particles, it is understood that it is the proper time that is being given. (As an aside, this shows that if we happen to know T and T_0, then γ can be obtained by just taking the ratio of T/T_0.)

We now want to consider the situation moving with the muon. First, what does the muon observe as far as Mt. Everest is concerned? From its point of view, Mt. Everest is moving upward with a speed very close to the speed of light. Also, since we are in the reference frame of the muon, the muon lives for its proper time of 2×10^{-6} s. What, then, does the muon (an observer moving with the muon) say about the height of Mt. Everest? We can again use d = vT to find the distance from the top to the bottom of Mt. Everest, with v = c and T = T_0. This gives d = cT_0 = (1.86×10^5) (2×10^{-6}) = 3.7×10^{-1} = .37 miles. In words, the observer who sees Mt. Everest moving observes it shortened (contracted) compared to an observer who is at rest with respect to Mt. Everest. The factor by which it is contracted is 5.5/.37 = 15. This, of course, is just the same relativistic factor γ we had above. This allows us to write the length contraction equation:

$$L = L_0/\gamma = L_0\sqrt{(1 - \beta^2)} \tag{12.16}$$

L_0 is the proper length (the length measured by an observer who is at rest with respect to that length). L is the contracted length (i.e., the length as measured by an observer who is in relative motion to the object whose length is being measured).

To make sure we understand the consequences of length contraction, let us consider our two observers both carrying identical meter sticks going by each other in their rockets with a relative speed close to c. Observer A holds out his meter stick so that observer B can measure it with hers. B is also holding out her stick so A can also measure her stick. What do both claim?

Since B is measuring A's stick, which is moving by her, she will claim that A's stick is no longer a meter long but has contracted. But from A's point of view, it is B's stick that is moving by him, so he must say that B's stick is contracted. Also, of course, each one says their own meter stick is exactly 1 m long.

It appears there is something very wrong here! How can both meter sticks be shorter than the other? *Shorter* is a comparative word. If one stick is shorter, English and logic tell us the other must be longer. We seem to have a paradox! How do we get out of it?

We have to think carefully about how we measure the length of an object that is moving by us. When we measure a length that is at rest with respect to us, we can line up one end with a mark on a ruler and then leisurely line up the other end with its appropriate mark. We could even take a coffee break before getting to the other end. But when an object is moving by, we do not have that luxury. If we used the same procedure as when the object was at rest, then by the time we got to the second end, the object would have moved and we would measure the length incorrectly. To avoid this, we must measure both ends *at the same time*. In other words, both ends must be measured *simultaneously*. Does this ring a bell?

OK, let us go back to A and B measuring each other's meter sticks. They now realize that in order to do the measurement correctly, they must measure both ends of the other meter stick simultaneously. But we have learned previously that two observers must disagree on whether events are simultaneous if they are in relative motion. So, A says to B, "I have done the correct procedure of measuring both ends of your meter stick at the same time and it is contracted." But B says to A, "I have watched you very carefully and I beg to differ; you have *not* measured both ends of my meter stick at the same time. No wonder you've measured the wrong length." Of course, when B measures the length of A's stick, we will get the same dialog.

We see there is no paradox. Each observer does indeed measure the other length as being shorter than his or her own. But each will claim the other has not done the measurement correctly because of the relativity of simultaneity.

We now understand why those three basic, ingrained assumptions are wrong. The relativity of simultaneity, time dilation, and length contraction certainly seem unusual, but remember, none of us have experienced the high speeds necessary to be able to notice these effects. You should note that while they seem to be three separate effects, they are all intertwined. There can be no length contraction without time dilation, and vice versa. Also, as we have just seen, we need to understand the relativity of simultaneity if we are to really understand length contraction.

12.8.3.1 Length and Lorentz Contraction

Before leaving our discussion of length contraction, there is one point that should be clarified. If we look at Equation 12.16 and remember that β is (u/c), then it appears that the length contraction equation is identical to the Lorentz contraction equation. So, was Lorentz correct after all? The answer is no. The meaning of the symbol u is very different in the two equations. In Lorentz's equation, u is the definite speed relative to the ether (which we now know does not exist). In Einstein's length contraction equation, u is the relative speed between any two observers. In looking at any equation, it is always important to make sure we understand the meaning of the symbols.

12.9   E = Mc² AND ALL THAT

Probably the best-known equation associated with the theory of relativity is $E = mc^2$. We have already introduced it at the end of Chapter 9, where it was used to express masses in MeV units. Before we discuss its full meaning, it will be useful to introduce a few other relativistic relations:

$$E = \gamma E_0 \qquad (12.17)$$

$$m = \gamma m_0 \qquad (12.18)$$

Equation 12.17 relates the energy as measured by two observers, while Equation 12.18 is the relation between masses. The 0 subscript, as usual, denotes the proper quantity. In Equation 12.18, m_0 is also known as the rest mass as well as the proper mass. These equations tell us that two observers will measure different energies and masses depending on their relative speed.

Either of these equations also explain why no object with mass can travel at the speed of light. If $u = c$, then $\beta = 1$, which would make $\gamma = 1/0$, which is infinite. Since there is not an infinite amount of energy or mass in the universe, then no massive object can ever attain the speed of light.

The relation $E = mc^2$ has two somewhat different meanings. If we consider a massive object that is at rest with respect to us, then we can write for its rest energy

$$E_0 = m_0 c^2$$

This tells us that an object with mass has energy. In other words, mass and energy are not separate quantities, but mass can be turned into energy and energy into mass. This, in essence, does away with the two separate laws of conservation of energy and conservation of mass and replaces them with the single law of conservation of energy. In fact, we used this in Chapter 10 when we calculated the kinetic energy available in beta decay. We should note here that in many situations, where there is essentially no conversion of mass into energy, using conservation of energy and mass separately is valid and usually much more convenient.

The second meaning comes about if we consider the case when a mass is not at rest relative to us. We then write

$$E = mc^2$$

And using Equation 12.17, we write

$$E = (\gamma m_0)c^2 = \gamma(m_0 c^2) = \gamma E_0$$

which, as you see, gets us back to Equation 12.18, which shows the consistency of these equations. The meaning of this is that the total energy of a moving object is proportional to its relativistic (moving) mass. Then, the kinetic energy, which is just the energy of motion, is the difference between its total energy and its rest energy:

$$\text{K.E.} = E - m_0 c^2 = mc^2 - m_0 c^2 \qquad (12.19)$$

It can be shown, but we will not do it here, that at speeds small compared to the speed of light, this expression reduces to the old familiar classical expression of K.E. $= (1/2) mv^2$.

Several examples of the importance of mass-energy conversion are listed below:

- In decays of nuclei and other subatomic particles (such as beta decay), where kinetic energy is released due to the fact that the masses of the final, decay products are less than the mass of the original particle.
- In a nuclear reactor, where energy is released when a heavy nucleus breaks up into lighter nuclei. This process is called fission.
- In the center of stars, where light nuclei combine with other light nuclei to form a heavier nucleus. But the mass of the heavier nucleus is less than the combined mass of the lighter nuclei. This is known as fusion, and it is the process by which all stars burn. Hopefully we will learn how to do this in a controlled way on earth, giving us an almost inexhaustible source of energy. It is a very difficult problem. Scientists have been working on it for over 40 years without success.
- In the process known as annihilation, where a particle and its antimatter partner collide and completely disappear. Their masses are converted totally into energy.
- In particle accelerators, where energy is imparted to the particles, such as electrons or protons. Initially the particles speed up, but as they get very close to the speed of light, the energy goes into increasing their mass, not their speed.

12.10   BACK TO ADDITION OF SPEEDS

In Section 12.5, we stated that Galileo's addition of speeds equation $v = v' + u$ cannot be correct since it leads to such nonsensical results. The correct equation must satisfy two criteria. At low speeds it must agree with Galileo's equation above, since we know that it works fine at normal everyday speeds. At speeds close to c, it must give the result that nothing can be measured as having a speed greater than c. The equation that Einstein derived does both. It is

$$v = (v' + u)/(1 + uv'/c^2) \tag{12.20}$$

Let us first consider this equation when u and v' are small compared to c. The term uv'/c^2 then has a value that is much smaller than 1 and can therefore be ignored. The denominator is essentially equal to 1. Equation 12.20 is then identical to Equation 12.1, the Galilean equation.

Let us go to the other extreme, $v' = c$, which corresponds to the case pictured in Figure 12.6. In fact, it was the discussion relating to that figure that led us to the nonsensical results if we insisted on using the Galilean equation. Now let us apply Equation 12.20 with $v' = c$:

$$v = (c + u)/(1 + uc/c^2) = (c + u)/(1 + u/c)$$

To get our final result, we will multiply the top and bottom of the last expression by c.

$$v = c(c + u)/c(1 + u/c) = c(c + u)/(c + u) = c$$

In words, Equation 12.20 gives us the result that the two observers, independent of their relative motion, both measure the speed of light as being the same. This, of course, is just the second postulate of relativity, and the correct equation must be consistent with it. If v′ and u have values less than but close to c, then v will be less than c also. Try it yourself.

12.11   THE CAR IN THE GARAGE PARADOX

There are several interesting and fun paradoxes that come about in relativity. We will discuss one of them that has to do with a car in a garage.

Let us consider a garage that has both a front and back door and a car that fits into the garage when the car is not moving with respect to the garage. As usual in relativity, we will have two observers: one at rest with respect to the car and one at rest with respect to the garage. The situation is shown in Figure 12.13.

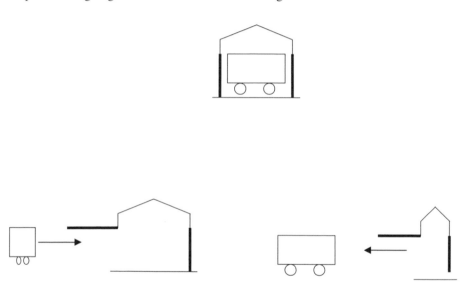

FIGURE 12.13   The car and the garage. In the top view, the car is at rest in the garage and fits into it. In the two bottom views, the car is approaching the garage at a speed close to the speed of light. The left view shows the situation as seen by an observer at rest with respect to the garage who sees the car length contracted. But from the point of view of the car observer (right view), the garage is approaching the car and is length contracted.

The garage observer has switches that allow him to open and close the doors very rapidly, so no damage should be done to the car or garage. In the case on the left, where the garage observer sees the length-contracted car approaching the garage, he can quite easily close the garage doors when the car is inside the garage and open them again before the car hits the far door. But as seen by the car observer, this appears impossible since that observer sees the garage length contracted so that it is much shorter than the car. In other words, it would appear that the car and the garage should be quite damaged. Now, while the observers might disagree about length, time, and other variables in relativity, they cannot disagree with certain facts. Both must agree that either there is damage or there is not. We seem to have a paradox.

Do you see how to resolve this? Try before reading further. (Hint: This is really very similar to something we have discussed above.)

Hopefully you recognized that the car-garage paradox is the same as the seeming paradox in length contraction we discussed in understanding how two observers could both claim the other's meter stick is shorter. The garage observer will claim that he both closed and opened the doors at the same time (i.e., simultaneously) so no damage is done. The car observer will indeed agree that no damage is done because the doors were not opened and closed simultaneously. She claims that the front door is opened first and the far is closed, and then the far door is opened and the front door closed, in the order shown in Figure 12.14. Thus, she claims the car gets through the garage with no damage.

FIGURE 12.14   Garage door opening-and-closing sequence as seen by the car observer.

12.12   THE TWIN PARADOX AND SPACE TRAVEL

The above is only one of several supposed paradoxes in relativity. Another is the well-known twin paradox. It has to do with the case of a set of twins. One stays on earth and the other goes off in a rocket with a speed close to the speed of light. The second twin returns to earth so that they can compare how much each has aged. The earth twin can claim since the rocket twin was in motion relative to her, the rocket twin will have aged less due to time dilation. In other words, all the rocket twin's clocks, including his heart, will have slowed down. On the other hand, the rocket twin can claim, from his point of view, the earth twin is the one in motion, so all of her clocks, including her heart, will have slowed down. Thus, they both claim the other twin has aged less. Certainly this appears to be a paradox—they both cannot have aged less than the other.

A correct analysis shows that the rocket twin has indeed aged less than his earth twin sister. In fact, if the rocket had a speed of 0.998c ($\gamma = 16$) and went off to Altair, a star 16 light years from earth, then returned, the earth twin would have aged 32 years while the rocket twin would have aged only 2 years. We will not go into a discussion of the explanation here (maybe your instructor will discuss it). Rather, it can be used to point out that this makes space travel possible. Altair, only 16 light years from earth, is a relatively close star. If we wanted to travel to another galaxy, we are then talking about distances of millions of light years.

The good news is that a rocket explorer, in principle, could visit another galaxy in a relatively short time (several years). The bad news is that everyone he knew back on earth would have died a million or more years ago. You could also look at this as a form of time travel in that you would be back at earth a million years into the future. Of course, you could not get back to the present.

At this point it might be interesting for you to consider the following question: How long would it take a light beam to travel across the universe, or any other distance for that matter? The answer may surprise you.

12.13 RELATIVITY AND YOU

You may be thinking at this point that this is all very interesting, and it may be even fun to contemplate some of these strange ideas, but how does it affect you? The odds are that in your lifetime, you, or anyone else for that matter, will never be able to achieve speeds anywhere near the speeds necessary to notice any relativistic effects. So there will be no effect on you at all. But in one sense, it can be argued that relativity may have already had a very profound effect on all of us.

It is generally accepted that mutations have been the driving mechanism in evolution. Mutations can be caused by copying errors in the genetic material, viruses, chemical pollutants, and naturally occurring radiation. It is safe to assume that if any one of these were absent, the rate of evolution would be affected.

Chemicals in the environment are a relatively new source on the evolutionary timescale since they have become important only after the industrial revolution. Most radiation, on the other hand, has been here since the earth was formed. It is due to several different sources: terrestrial sources in the ground and air, internal sources due to radioactive atoms in our body, and cosmic radiation impinging on the earth from the sun and other astronomical sources. The vast majority of cosmic rays at sea level are the muons we discussed earlier. They are passing through us several times each second. As with the other sources of radiation, they ionize the water molecules in our body to create OH^- radicals, which are very reactive and can change our DNA. It is this effect on our DNA that causes the mutations. So, what does this have to do with relativity?

Remember we discussed earlier that muons have a lifetime of 2×10^{-6} s. Living that long, they would travel only about 1/3 of a mile or less before decaying. Since the muons are produced near the top of the atmosphere, none would reach the surface of the earth. Hence, they could not cause any mutations, and *Homo sapiens* might still be living in cold caves, not yet having been able to figure out how to make fire. But due to time dilation, the muons live longer and have been able to reach sea level and cause some of the mutations that have allowed us to evolve into what we are today. So, if it were not for relativistic time dilation, you may not be in your comfortable room reading this, but instead could be wrapped in some animal skin shivering in a cold, dark, dank cave, or, worse still, just crawling out of the primordial sea.

12.14 END-OF-CHAPTER GUIDE TO KEY IDEAS

- What is the statement of Galilean relativity? What are the important words in this statement and what do they mean?
- What is the Galilean transformation of speeds and what does it mean?
- What was Maxwell's discovery and how is it related to the ether?
- What determines the measured speed of a wave? Of a particle?
- Why was the Michelson Morley experiment performed?
- What was the basic idea of the experiment?
- What was observed in the Michelson Morley experiment?
- What were the possible explanations for the observation in the Michelson Morley experiment?
- Which explanation do we accept today. Why were the others rejected?
- Why does the accepted explanation initially sound so crazy?
- What are the three assumptions we make when two observers move by each other?
- What are the postulates of special relativity? Which was an old postulate and which was new?
- What is the medium for light?
- What is the relativity of simultaneity?
- What is time dilation?
- What is length contraction?
- What are the equations for time dilation and length contraction? What are the meanings of the various terms in each equation?
- Are length contraction and the Lorentz contraction the same thing?
- What is the meaning of the equation $E = mc^2$?
- What is the new addition of speeds equation? How does it differ from the Galilean equation?
- What is the "car in the garage" paradox? What is the solution to the supposed paradox?
- Is space travel possible? What is the downside?
- How has relativity possibly had an effect on evolution?

QUESTIONS/PROBLEMS

1. What is a frame of reference? Give three different examples.
2. How fast are you moving as you read this?
3. Are you at rest as you read this?
4. What does it mean to say "I am at rest"?
5. What is meant by an absolute frame of reference?
6. Mary moves away from Bob at half the speed of light. Bob shines a searchlight toward Mary. What does Galilean relativity predict for the speed of the light beam that Mary measures?
7. According to Galilean relativity, does every observer, independent of his or her motion, measure the same speed for light?

8. Mary is moving at a speed of 2 m/s while Bob, who is standing at the side of the road, throws a ball in the same direction as Mary is moving, at a speed of 8 m/s.
   a. What is the ball's speed and direction (relative to Mary's motion) as measured by Bob?
   b. What if Bob had instead thrown the ball in the opposite direction at 8m/s?
9. Mary moves at a speed of 2 m/s. She throws a ball in the same direction she is moving, at a speed of 8 m/s.
   a. If Bob is standing at the side of the road, what does Bob measure for the speed and direction of the ball?
   b. What if Mary, instead, had thrown the ball at 8 m/s in the opposite direction she was moving?
10. Bob is on a rocket moving toward you at 0.9c. He sends a light beam toward Mary on earth.
   a. How fast does Bob see the beam move away from him? Answer for both Galilean and special relativity.
   b. What does Mary measure for the speed of the light beam?
   Answer for both Galilean and special relativity.
11. a. Is it possible (i.e., consistent with the principles of physics) for a person in a rocket to travel toward earth with a speed of 0.99c?
   b. What would be the light speed as measured by an earth-bound observer? Answer for both Galilean and special relativity.
   c. If an earth-bound light was projected toward the rocket, what speed would the rocket observer measure? Answer for both Galilean and special relativity.
12. The statements below have to do with an observer measuring the speed of a projectile or a wave. Indicate whether each statement is true or false. If false, correct the statement or tell why you think it is false.
   a. The measured speed of a projectile does not depend on the relative speed of source (reference frame of the launcher of the projectile) and observer.
   b. The measured speed of a projectile does not depend on the speed of the source.
   c. The measured speed of a projectile does depend on the relative speed of the observer.
   d. The measured speed of a wave does not depend on the relative speed of the source and observer.
   e. The measured speed of a wave depends only on the properties of the medium.
   f. The measured speed of a wave does not depend on the motion of the source.
   g. The measured speed of a wave depends only on the motion of the observer through the medium.
   h. If a source of a wave and an observer are moving through a medium with the same speed, the measured speed of the wave is independent of that speed.
   i. If a source of a projectile and an observer are moving with the same speed, the measured speed of the projectile is independent of that speed.

13. A boy is on a train that has a speed of 50 mph relative to the ground. He has a ball that he throws at 20 mph relative to the train in the same direction the train is moving.
    a. Relative to an observer on the ground, how fast is the ball moving?
    b. If there is an observer on another train, moving at 50 mph in the same direction as the boy's train, how fast does this observer see the ball moving?
    c. If the boy faces the front of the train and blows a trumpet (the speed of sound in still air is 600 mph), what speed will the ground observer measure for the sound waves from the trumpet?
    d. If there is an observer on the boy's train and he is standing in front of the boy, what will that observer measure for the speed of the trumpet waves?
    Note: For (c) and (d), assume the train cars are completely open.
14. a. What is the ether?
    b. What does ether theory predict about what determines the measured speed of light?
    c. What are possible ways of detecting the ether?
15. a. Describe a very simple method to measure the speed of the earth through the ether.
    b. Why doesn't this method work?
16. a. Why was the Michelson Morley experiment performed?
    b. Describe the idea as to how the desired result would be achieved.
    c. What was the expected result?
    d. What was observed?
17. Derive Equation 12.3 by carefully going through the steps yourself.
18. Derive Equation 12.8 by carefully going through the steps yourself.
19. In the analogy between the Michelson Morley experiment and boats in a river, what are the boats and the river analogous to?
20. List two possible *natural* explanations for the observation in the Michelson Morley experiment. Discuss why neither was valid.
21. Describe Lorentz's idea. Why would that explain the result of the Michelson Morley experiment?
22. How does the fact that the speed of light is always the same explain the result of the Michelson Morley experiment?
23. Why did the explanation that the speed of light is always the same seem like such a crazy idea at the time?
24. State the two postulates of special relativity. Which is based on a symmetry principle, and why is this so?
25. When two observers in relative motion measure the speed of a third object, what assumptions do we normally make about each observer's measurements?
26. Carefully define simultaneous events.
27. State the principle of the relativity of simultaneity.
28. Under what conditions will all observers agree that two events are simultaneous?

29. Discuss all the relevant reasons why observers A and B in Figure 12.9 will not agree that events 1 and 2 are simultaneous.

30. Define *time dilation*.

31. Can you explain time dilation to a nonphysicist friend?

32. What is special about the light clock?

33. Is the light clock better than other clocks?

34. Define *proper time*.

35. Calculate the value of γ for values of $\beta = 0.1, 0.2, 0.3, 0.5, 0.6, 0.8, 0.9, 0.95, 0.99$, and 0.999, and make a graph (using graph paper or a plotting program) of γ (dependent variable) versus β (independent variable).

36. What are the minimum and maximum allowed values of β and γ?

37. If an observer moving at 0.6c with respect to you reports that an event on her rocket took 10 min according to her clock, how much time would you claim elapsed on your clock?

38. If an observer moving at 0.6c with respect to you reports that an event on your rocket took 10 min according to her clock, how much time would you claim elapsed on your clock?

39. Consider the same two rockets as in the previous problems. You measure the length of an object in your rocket to be 10 m. What does the observer in the other rocket measure for the length of the object?

40. Again, consider the same two rockets as in the previous problems. You measure the length of an object in the other rocket to be 10 m. What does the observer in the other rocket measure for the of the object?

41. When at rest, muons (μ) have a half-life of 2×10^{-6} s. One thousand muons at a particle accelerator are moving at a very high speed such that after 200×10^{-6} s, 250 muons have not decayed.

    a. According to an observer at rest with respect to the accelerator, what is the half-life of these muons?

    b. What is the value of γ for these muons?

    c. For an observer moving with the muons, how much time did it take for the number of muons to reach 250?

    d. How many meters did the muons travel in the accelerator lab before 500 of them would have decayed?

42. A rocket observer and a platform observer both measure the length of the same stick. The rocket observer measures it to be 80 cm, while the platform observer measures it to be 20 cm.

    a. One of the observers is at rest with respect to the stick. Which one?

    b. What is the value of γ for this case?

    c. An event occurs in the rocket frame such that the platform observer claims the time for the event is 20 s. What does the rocket observer claim for that time interval?

    d. Briefly discuss how the rocket observer explains the fact that the platform observer measures a different length for the stick.

43. How is it logically possible for two different observers measuring the length of the same object to both say that the object is shorter than the other observer claims?

44. A clock on a rocket passes you with a speed of $v = 0.98c$.
    a. What is the value of γ?
    b. According to your watch, how long does it take the moving clock's second hand to make one complete revolution?
    c. What length do you measure for a meter stick that is on the rocket?
45. The star α-Centuri is 4 light years from earth.
    a. To an observer on earth, how much time does it take for a light beam to reach α-Centuri?
    b. The rocket travels at $v = 0.98c$ relative to earth. To an observer on the rocket, what does she measure for the earth, α-Centuri distance?
    c. How much time does she claim it takes to go from earth to α-Centuri? (Be a bit careful here.)
    d. If an observer could travel at the speed of light, how much time would she claim it took to reach α-Centuri? (This is a good one!)
46. How fast must a rocket move past you if you the length of the rocket is half of its proper length?
47. A meter stick with a rest mass of 1 kg moves past you such that you measure its mass as 2 kg. What length do you measure for the meter stick?
48. Consider a rocket and platform approaching each other at a relative speed of 90% the speed of light. A projectile on the rocket is launched toward the platform at a speed relative to the rocket at 70% the speed of light. What speed does the platform observer measure for the projectile?

13 Quantum Mechanics

13.1 INTRODUCTION

What is usually called modern physics is based on two theories: relativity, which we have just finished discussing, and quantum mechanics (also called quantum theory). While some of the ideas of relativity, such as time dilation and length contraction, at first seemed rather strange, we saw that they followed logically from the two postulates of relativity. We will see below that the ideas of quantum mechanics are also strange. Most physicists would agree that they are much stranger than relativity. In fact, we will see that there are no satisfactory explanations. To quote two Nobel Prize laureates, Murray Gell-Mann and Richard Feynman, "Nobody really understands quantum theory." Despite this, it is one of the best-tested theories that we have. In some cases, the agreement between the theoretical prediction and experiment is better than 1 part in a million. There are many useful devices that are based on quantum mechanics. They include anything that uses microelectronics, such as a cell phone, DVD player, or computer, nuclear reactors, and medical imaging devices such as MRI and PET.

13.2 MAX PLANCK AND THE BEGINNINGS OF QUANTUM THEORY

At the end of the nineteenth century, there were concerns among physicists about understanding how a hot object, such as a burning log in a fireplace, emitted radiation. This was known as the problem of black body radiation. Basically any object with a temperature emits electromagnetic radiation. For humans, the frequency of that radiation is mostly in the infrared range. As you are sitting and reading this, you are radiating power equivalent to that of a 100-watt lightbulb. If you were jumping up and down, as you possibly might in a rock concert, you would be radiating even more energy. That is why concert halls can get quite hot. Everyone is surrounded by 100-watt (or more) lightbulbs.

Without going into details, the predicted frequency spectrum of light emitted from a hot object did not agree with the experimentally measured spectrum. Worse still, the theory, based on the classical ideas of both mechanics and electromagnetism, predicted that such a hot object should emit an infinite amount of energy—clearly impossible. In 1900, Max Planck, a German physicist, introduced a new, *ad hoc* idea that predicted the correct spectrum. Before Planck, it was logically assumed that the energy of the particles in the hot material that were emitting the light could have a continuous distribution. In other words, any value of energy was allowed. Planck's idea was that this was not the case. Only discrete values of energy were permitted, and these values were given by the simple equation

$$E = n(hf) \tag{13.1}$$

where E is the energy of the radiating particle, f is the frequency of oscillation of the particle, h is a constant of proportionality to make the units correct, and n is an integer (n = 1, 2, 3, etc.). Because of the integral values of n, Equation 13.1 says that the energies could not have any value but only certain values that were multiples of the basic unit of energy (hf). For instance, E could not have values such as 1.5(hf) or 2.73(hf). Also, there is a smallest, minimum energy given by (hf). In other words, the particles could only have specific units or bundles of energy. The Latin word for "bundle" is *quantum*—hence the name *quantum theory*. It is common to talk about the fact that the energy is *quantized*. This was certainly a very strange idea with no basis for suggesting it, except that it worked in predicting the correct spectrum of light. By comparing his theory with the experimental data, Planck could calculate the value of the proportionality constant h. It has the very small value of 6.6×10^{-34} Joule-s and is now known as Planck's constant (Joule-s are the units needed to convert frequency, which has units of 1/time, to energy, which has the units of Joules in the MKS system). Planck did not like this idea at all since there seemed to be no classical foundation for it and tried to find some other explanation.

13.3   THE PHOTOELECTRIC EFFECT

We now jump five years in time to 1905. Remember that was the year that Einstein published his relativity paper. You may remember that he also published four other papers in that year. One of these, on what is known as the photoelectric effect, presented ideas that were much more controversial than those of relativity because it suggested that light was not a wave phenomenon but instead was composed of particles. This surely had to be wrong given the evidence we discussed in Chapter 11 for the interference, diffraction, and polarization of light—all being wave properties that particles cannot exhibit.

The photoelectric effect was discovered in 1887 by Heinrich Hertz, who was trying to verify Maxwell's theory about electromagnetic waves (the unit of frequency, Hz, is named after him). Hertz observed an electromagnetic wave by having it produce a spark between two separated metal electrodes. He noted that he could obtain a larger spark if he shined light on the electrodes. Shortly after, it was determined that the light caused electrons to be emitted from the metal of the electrodes. In 1902, Philipp Lenard showed that that number of electrons emitted depended on the intensity of the incident light (which was expected from the wave theory of light), but the kinetic energy of the electrons did not depend on the intensity. In fact, it seemed to depend more on the color (frequency) of the light (which was not expected according to the wave theory of light). Now, the photoelectric effect is the basis for light meters and photovoltaic cells, to name a few applications.

Basically, Einstein said that light acted as particles in knocking the electrons from the metal. The properties of these particles, now called photons, were based on Planck's idea that was published five years earlier. Namely, the energy of a photon is given by $E = hf$. It should be noted that while the energy relation is the same as

Planck's, the meaning is quite different. Planck had said that the particles in the metal, the electrons, had energies that were quantized. Einstein said that light itself, wherever it came from, acted as if it came in bundles of energy-like particles.

In his paper, he made specific predictions of how the emitted electrons would behave depending on the nature of the incident light. These predictions were very different from what would be predicted if light was made up of waves. Let us look at these predictions for both waves and particles in the table below.

Property of Emitted Electron	Wave Prediction	Particle Prediction
Energy	Depends on intensity of the light	Depends on frequency of the light
Time for emission	Depends on intensity	Instantaneous
Threshold frequency	None	Yes, dependent on metal
Number/second	Depends on intensity	Depends on intensity

To understand these predictions, let us try to visualize what is happening. Our picture is that inside the metal the electrons are freely moving but cannot escape because there are atomic forces holding the electrons in the metal. The amount of energy necessary to overcome these forces, and thus free the electron, is known as the work function of the metal. Different metals have different work functions. When struck by the incident light, the electron could gain enough energy to get free if there is enough transfer of energy from the light to the electron.

For waves such as light, the intensity is related to the energy of the wave. So the energy of the ejected electron should depend on the intensity of the incoming wave that strikes it. Also, the time it takes to free the electron from the metal should depend on the energy of the wave. The more energetic the wave, the less time it should take to transfer enough energy to the struck electron to free it. Since the incoming energy depends on the intensity of the light, there should be no dependence on the frequency. Finally, the higher the intensity of the light, the more electrons that should be knocked out.

For Einstein's particle picture, the photon collides directly with an electron, transferring all of its energy to it. It is like a billiard ball collision, where all of the energy of the incoming ball is transferred to the struck ball. Since the photon's energy is hf, the kinetic energy of the struck electron has to be proportional to the frequency. With a direct collision, there is an instantaneous transfer of energy, so there is no time dependence on intensity. The reason for a threshold (minimum) frequency is that if the photon does not have enough energy to free the electron from the metal, then no electrons can be knocked out. Again, since the photon energy depends on frequency, if the frequency is not high enough, there is not enough energy to overcome the work function. Einstein's paper contained a simple equation for the kinetic energy of the emitted electron:

$$K.E. = hf - W \qquad (13.2)$$

where W is the work function for the metal. We see that if the photon energy, hf, is less than W, no electrons can be ejected since kinetic energy cannot be negative.

Finally, in the photon picture, the intensity of light is directly proportional to the number of photons. Thus, the more intense the light, the more photons available to strike electrons and knock them out of the metal.

As the table above shows, the two models for light make very different predictions for three of the four properties listed. This, again, is just the type of situation physicists love since an experiment should be able to easily distinguish between the two. It was not until 1916 that the American physicist Robert Millikan was able to do the definitive experiment, which confirmed Einstein's predictions. Five years later, in 1921, Einstein was awarded the Nobel Prize for his 1905 paper on the photoelectric effect.

At this point, let us recap. In Chapter 11, we showed that due to the evidence of interference, diffraction, and polarization, light had to be a wave phenomenon. Now we see that the photoelectric effect can only be understood if light is a particle phenomenon. Confusing? It sure is.

Before delving further into some of the other rather strange aspects of quantum theory, let us pause and make sure we understand the audacity of Einstein's prediction. Remember, in 1905 the wave nature of light was an established "fact," rooted in both experimental observation and Maxwell's well-accepted theory. Yet, based on Lenard's rather flimsy observation 3 years earlier, Einstein decides light must act as composed of particles. In doing this, he also has to take Planck's strange and not really believed *ad hoc* idea about the relation between energy and frequency and then apply it not to electrons, as Planck had done, but to light. In addition, Einstein was not some highly respected professor, but a rather lowly patent examiner. What was it in Einstein's psyche that allowed him to think so far out of the box and also go against established thinking? This, of course, is a question for psychologists, not physicists. But it is certainly interesting to think about.

13.4 THE BOHR ATOM

In Chapter 9, we stated that after the results of the Rutherford experiment, the Danish physicist Neils Bohr proposed some revolutionary ideas that allowed him to predict the correct spectrum of light emitted by the hydrogen atom. Given our discussion above, we are now in the position to start to be able to appreciate his very novel ideas.

Before going into the details of the Bohr model, let us remember that the Rutherford experiment had shown that the plum pudding model could not be correct. This left the planetary model as the only other possible picture for the atom. But this model had the serious problem that the physics known at the time predicted that the electron in its orbit would radiate away its energy in a time of about 10^{-6} s. Thus, the atom could only exist for this time. Also, the frequency of light emitted from the electron would be continuously increasing as the electron would spiral into the nucleus. This was in contradiction with the observed discrete spectrum of light emitted by the atom.

What Bohr did was to say that since we know atoms exist, and exist for very long times (comparable to the age of the universe in some cases), some parts of classical physics cannot be correct, and new theories are needed. For the atom, he put his new ideas in the form of rules, some of which were based on the ideas of Planck and Einstein. These rules are as follows:

1. Electrons are only allowed to exist in certain discrete circular orbits with corresponding discrete energies.
2. While in any of these orbits, the electron does not radiate energy.
3. The electrons radiate energy only when making a transition from a higher to a lower energy level.
4. The energy of the emitted photon is just the energy difference between the two states, and the frequency of that emitted radiation is related to the energy, E, of the photon by the equation $E = hf$ (the Einstein relation for the photoelectric effect).
5. The condition that determines the nature of the discrete orbit has to do with the angular momentum, L, of the orbit. The angular momentum can only have discrete (quantized) values given by $L = n(h/2\pi)$.

Before going into these rules in some detail, we need to talk about angular momentum, since we have not mentioned it previously. Momentum, mv, which we have discussed before, is also known as linear momentum because it usually has to do with straight-line motion. When an object moves in a circle, the angular momentum is usually the more appropriate quantity. It is closely related to the linear momentum and is given by the expression $L = mvr = pr$, where r is the radius of the circle. Just like momentum, angular momentum can also be conserved under the right circumstances. It is an important variable in describing circular motion. For our purposes, that is all we have to know about it. We bring it up here because, as rule 5 indicates, it has significance in understanding the electron's motion in its circular orbit in the atom. Namely, it is the relevant variable that is quantized. We cannot go into the details here as to why Bohr chose to quantize angular momentum. But as a hint, while we stated above that Planck's constant, h, has the units of Joule-s ([energy] × [time]), it also has the units of angular momentum. You should be able to show this for yourself.

Let us now take a closer look at all of Bohr's rules:

Rule 1: Planck's original idea that energy is quantized.
Rule 2: New, *ad hoc* idea. Even though it goes entirely against classical physics, Bohr knew that the atom did not ordinarily radiate. Rule 2 is just a statement of this experimental fact.
Rule 3: States that when the electron is excited to a higher energy state it will go to a lower energy state due to the energy prime directive mentioned in Section 9.6. But it can only go to one of the allowed, discrete energy levels according to rule 1. When the electron goes from a higher to lower energy state, the loss of energy appears as radiation. It was certainly known that the atom could be caused to radiate when excited by some outside excitation. This is how atomic spectra are produced.

Rule 4: A statement of conservation of energy coupled with Einstein's hypothesis in his photoelectric effect paper that light has an energy given by $E = hf$.

Rule 5: Defines what property of the motion (angular momentum) determines the discrete levels.

When Bohr combined these rules, he was able to derive the value of the energy of each allowed level. Knowing the energy, he then knew the energy difference between each level. Using rule 4, he could then derive the frequency of the emitted radiation. These frequencies had to be discrete, as was one of the heretofore mysterious observations. Even more importantly, his predicted frequencies agreed with the already known measured frequencies. Obviously, Bohr was on the right track even though his *ad hoc* rules were not understood at the time. It would take about another 10 years before his angular momentum quantization rule 5 would be justified by a French PhD student.

13.5   DE BROGLIE WAVES

In 1924, Louis de Broglie (pronounced de Broy) completed his thesis for the PhD degree at the University of Paris. In that thesis, he suggested that there should be a symmetry between particles and waves. If, as Einstein had suggested (and by that time had been verified), waves (light) could exhibit particle properties, why couldn't particles have wave properties? In other words, there should be a wave-particle duality.

As Einstein had given the relation between the wave property (frequency) and particle property (energy) as $E = hf$, de Broglie hypothesized the relation between the wave property (wavelength, λ) and particle property (momentum, p) as being given by

$$\lambda = h/p \qquad (13.3)$$

where λ is now known as the de Broglie wavelength.

This was indeed a bold suggestion. In many ways it was much stranger than Planck's quantum idea or Einstein's photon idea. After all, we observe all sorts of particles (pieces of matter) every day and they certainly do not seem to exhibit any type of wave properties. The question, then, is if there is any validity to de Broglie's idea, how could it be tested?

In Chapter 11, we discussed important differences between particles and waves. One of the definitive tests for something being a wave was whether it exhibited interference or diffraction. Several years after the publication of de Broglie's thesis, it was demonstrated that electrons indeed formed interference patterns when shot through the equivalent of two slits. In these experiments, the slits had to be the regular spacing between atoms in a crystal. The first experiment was performed in 1927 by Davisson and Germer. In 1928, G. P. Thomson also demonstrated the interference of electrons. It is ironic that G. P. Thomson shared the Nobel Prize with Davisson and Germer for showing that electrons had a *wave*-like nature, while his father, J. J. Thomson, received the Nobel Prize for the discovery of a new *particle*, the electron.

At this point, you may be asking: So, if particles can behave as waves, why don't we see this behavior with billiard balls, baseballs, bowling balls, and all other familiar objects? To answer this, let us first calculate the de Broglie wavelength of some common object such as a billiard ball rolling on a table. A billiard ball has a mass of about 1 kg and a typical speed is about 1 m/s. Equation 13.3 gives:

$$\lambda = h/mv = 6.6 \times 10^{-34}/(1)(1) = 6.6 \times 10^{-34} \text{ m}$$

This is an extremely small wavelength, much smaller than the size of an atomic nucleus, which is the order of 10^{-15} m.

Back in Chapter 11, we discussed that in order for the wave nature (e.g., diffraction) to be observable, the wavelength of the wave had to be comparable to or larger than the object the wave was hitting. So, we see that for ordinary-sized objects the wave nature will never be observable. On the other hand, let us now calculate the wavelength of an electron ($m = 9.1 \times 10^{-31}$ kg) with a speed of 10^6 m/s.

$$\lambda = 6.6 \times 10^{-34}/(9.1 \times 10^{-31})(10^6) = 7 \times 10^{-10} \text{ m}$$

This is the order of atomic dimensions. Thus, wave properties such as diffraction and interference could be observed when electrons scatter off atoms. That is why both the experiments mentioned above had to use atomic spacings to detect the electron's wave nature.

The above discussion suggests something very interesting. If you remember in Section 11.5, we discussed the fact that the best optical microscope could not be used to ever see an atom. The reason was that the typical wavelength of visible light is about 1,000 times larger than the size of an atom. With such a large wavelength compared to the size of the atom, the light waves would completely diffract around the atom, and hence the atom would have no effect on the light. If an object has no effect on the wave, then there is no way of knowing the object is even there. But if we used electrons whose de Broglie wavelength could be made comparable to atomic dimensions, could they be used to observe atoms? The answer is yes. In fact, this is the basis for the electron microscope, where the wave nature of electrons is used to be able to observe objects whose size is much smaller than the wavelength of visible light. While the subatomic details of an atom have not been observed with electron microscopes, groups of atoms have been observed.

We can also use de Broglie's idea to shed new light on Bohr's rule 5, which we discussed above. Bohr based this assumption on classical ideas, but we can now use the de Broglie wavelength to give a quantum justification. If the electron has a wave nature, how does that manifest itself in its orbit in the atom? (Here we are not considering interference or diffraction.) The quantum answer is that since the electron moves in a circular orbit, the number of de Broglie wavelengths that fit into the circular orbit must be an integral number (i.e., quantized). Figure 13.1 shows one such configuration.

Let us see how this leads to Bohr's rule 5. We see from Figure 13.1 that a full number of wavelengths must exactly fit around the circumference of the circle. Thus, we can write

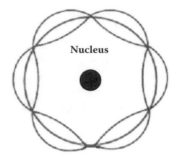

FIGURE 13.1   **(See color insert following page 102.)** de Broglie wavelengths in the circular electron orbit in the atom. This particular configuration corresponds to three integral number of wavelengths (n = 3).

$$n\lambda = 2\pi r \qquad n = 1, 2, 3, \dots \tag{13.4}$$

But according to de Broglie, $\lambda = h/mv$. Substituting this into Equation 13.4 gives

$$n(h/mv) = 2\pi r \tag{13.5}$$

Cross-multiplying and bringing mvr to the left side, we can write

$$mvr = L = n(h/2\pi)$$

which is Bohr's rule 5 quantizing angular momentum.

13.6   TIME TO STOP AND CATCH OUR BREATH

This would be a good place to stop and review what we have learned so far. Basically three things:

1. There are certain quantities, like energy and angular momentum, that cannot just have any value but are integral multiples of a number known as Planck's constant. We say such quantities are *quantized.*
2. Light, which was thought of as being a wave, has particle properties.
3. Particles, like electrons, have wave properties.

These all seem to be rather mysterious. Why don't we observe them in our everyday life? Basically because these effects only show up on the microscopic, atomic, and subatomic scales. So while bowling balls do not spread out and wiggle around like waves, electrons do. But there are circumstances where some of these effects can be indirectly observed on our macroscopic scale. For instance, the photoelectric effect is the basis for light meters in your cameras, or for electric-eye doors, or for solar cells to create electrical energy. Also, while we cannot see atoms, we can observe atomic spectra, which because of their discreteness show that atomic energy levels are quantized.

You might be asking at this point: Why do these effects only appear on the microscopic scale? It is because Planck's constant is so very small (6×10^{-34}). If it were a good deal larger, then these strange effects would be observable in our ordinary, everyday life. It would indeed be a very weird world we would be living in with all sorts of supposedly solid objects behaving as water waves, for example.

How do we understand how particles can sometimes behave with wave properties, and waves can sometimes behave with particle properties? It certainly sounds very strange. A good deal of the difficulty stems from our desire to categorize something as being *either* a wave or a particle. Our everyday experience, along with our discussion in Chapter 11, tells us that waves and particles are very different entities. But what quantum mechanics has taught us is that, at least on the microscopic scale, objects have *both* wave and particle properties. This is known as wave-particle duality, which we now believe is intrinsic in nature. Much of what we will be discussing below is based on this wave-particle duality.

13.7   THE HEISENBERG UNCERTAINTY PRINCIPLE

In 1927, the German physicist Werner Heisenberg, while pondering aspects of wave-particle duality, postulated the now famous relation that bears his name. It is one of the most important ideas in quantum mechanics. It has a dual meaning, but many times only one of those meanings is discussed. It also represents several different relations, depending what variables are of interest. One of these important relations is

$$\Delta x \Delta p \geq h/2\pi \tag{13.6}$$

The Δ symbol means uncertainty in this case. Thus, Equation 13.6 states that the uncertainty in position (x) times the uncertainty in momentum (p = mv) must be greater than or, at best, equal to Planck's constant (h) divided by the number 2π. Another way of saying this is that the product of these uncertainties cannot be less than the quantity ($h/2\pi$). Note that the value of this quantity is very close to 10^{-34}, a very small number indeed. Because of this, the uncertainty principle has no direct relevance in our day-to-day macroscopic world. On the other hand, as we shall see, it is very relevant on the microscopic scale.

The usual interpretation of Equation 13.6 has to do with inaccuracies when doing a measurement. If one were to try to measure the position and momentum of an electron, for example, at the same time, then the measurement uncertainties of those two variables could not be determined better than that given by Equation 13.6. This is just not a problem of having inaccurate equipment; rather, it is inherent in nature.

Why does wave-particle duality have anything to do with this? In order to answer this, let us consider trying to see an electron with light. The word *light* here means electromagnetic radiation with a wavelength small enough to be able to see the electron. We will need a very short wavelength to be able to see the electron due to the wave property of diffraction (remember the discussion in Chapter 11). The electron is very small (much smaller than a proton), so we have to use very short-wavelength light. What's the matter with that? Well, short wavelength means high frequency (also remember the relation $\lambda f = c$, or $f = c/\lambda$). But Einstein has told us that light also has a particle (photon) aspect where $E = hf$. So high frequency means a high-energy photon. If an electron gets struck by a high-energy particle, it will be given a large kick, so that right afterward, both its position and velocity (hence its momentum) will no longer be known very well. A more quantitative analysis shows that the smallest uncertainties of the position and momentum are given by Equation 13.6. If we turn the argument around by trying to use a low-energy photon that will not disturb the electron, then the low frequency of this photon means it will have a very large wavelength. But a large wavelength means that because of diffraction we will not be able to locate the electron very well, so its position will be very inaccurate. Either way, the combined wave-particle duality forces limits on how well we can do certain measurements.

As we have seen, an important aspect of measurement theory is that if we want to observe something, we must interact with the object we are trying to observe. Whatever means we use for the observation, this interaction must disturb the object, and hence affect the very thing we wish to observe. On the macroscopic scale, this interaction is usually so small it has no real effect. But on the microscopic scale, it can have a very significant effect.

Einstein was not very happy with this situation and proposed clever thought experiments to try to get around the limits imposed by the uncertainty principle. These are known as gedanken experiments. *Gedanken* is the German word for "thought." They were experiments that could not necessarily be performed due to technology limits but did not violate any known laws of physics. Einstein usually presented them in the morning at important physics conferences. By that evening, Niels Bohr would have a solution showing why Einstein's gedanken experiment would not violate the uncertainty principle. Once Bohr even used Einstein's own relativity theory to prove Einstein wrong.

So far, we have discussed the uncertainty principle in terms of inaccuracies that occur in the measurement of certain variables (e.g., position and momentum). This is the usual interpretation that is given. But there is another interpretation that has nothing to do with observation and measurements, but has to do with intrinsic uncertainties or fluctuations in nature. To discuss this, let us look at another form of the uncertainty principle.

$$\Delta t \Delta E \geq h/2\pi \tag{13.7}$$

This is the time (t), energy (E) form. It too can be discussed in terms of how well the energy can be measured in a given time interval, but we will use this to consider one of the strangest aspects of quantum mechanics.

In Chapter 8, we discussed the conservation of energy as one of the important ideas in physics. But one interpretation of Equation 13.7 is that for very short time intervals, energy does not have to be conserved. Again, given the very small value of Planck's constant, this energy nonconservation is minute on the macroscopic scale. On the other hand, on the microscopic scale, this has important and measurable consequences. We believe it is intrinsic in nature. On the atomic and subatomic scales, there are continual quantum energy fluctuations occurring with subatomic particles popping in and out of existence, consistent with Equation 13.7. You might ask at this point: "Where did these particles come from? I thought we are talking about energy." But remember, from our old friend $E = mc^2$, matter (mass) is just another form of energy.

We believe that this effect is continually happening everywhere. We call this the fluctuation of the vacuum, since even in what we may think is a perfect vacuum, particles appear and then, in a very short time interval, disappear again. How much time are we talking about? Well, for the case of an electron and antielectron positron "popping" into existence (we need both to conserve charge), the time interval until they disappear is about 10^{-21} s. At this point you have the right to be very skeptical—how can anything be measured in such a short time interval? If we cannot measure it, isn't this science fiction and not science?

Hopefully you have seen several times by now that many of our "measurements" are not direct, but yet certainly valid. While we cannot directly observe these fluctuations over such small time intervals, we can measure their effects. For instance, in the atom, when an electron-positron fluctuation occurs, the presence of the charges associated with these particles slightly modifies the electric field that the atomic electrons experience. This, in turn, modifies the energy levels of these atomic electrons. Remember, as we saw above in our discussion of the Bohr atom, the atomic spectra that we observe depend on the energy levels of the atom. Thus, if the energy levels of the atom are modified by these fluctuations, then the atomic spectra will also be modified. In fact, this modification or shift in the energy levels can be calculated and have indeed been measured. The measurements agree very well with the prediction based on our understanding of the fluctuations of the vacuum.

Another example of the effect of energy fluctuations is alpha decay. According to energy conservation, an alpha particle should never be able to escape from a nucleus. But as we discussed in Chapter 10, alpha decay is one of the three forms of nuclear decay. Not only do these fluctuations allow for alpha decay to occur, but we can also use the theory to predict alpha decay half-lives. Again, the measured half-lives confirm the predictions. In the case of alpha decay, this effect is known as tunneling since the alpha particle seems to be able to tunnel through a barrier, which would violate classical conservation of energy. Some microelectronic devices are also based on tunneling.

If you think things are getting pretty weird, it should be pointed out that there are some theories that have our entire universe starting from a quantum fluctuation of the vacuum. Of course, there will probably never be any experimental verification of this, but it is an intriguing idea.

13.8   THE SCHRODINGER EQUATION: AN EQUATION FOR THE WAVES

If particles have a wave nature, this leads to several questions. What exactly is this wave (i.e., what is "waving")? What determines how this wave moves? In 1926, 2 years after de Broglie's hypothesis, Erwin Schrodinger produced an equation that described how the wave should propagate. It was shortly afterward that Max Born gave an interpretation of what was waving.

The Schrodinger equation can be considered the quantum equivalent of Newton's second law. It is the forces acting on a system that determine the motion of the wave, in a similar manner that forces determine the acceleration of a particle. To be accurate, the equation is the quantum, mathematical statement of the conservation of energy. But if you remember the discussion in Chapter 8, the potential energy depends on what particular forces are acting. The Schrodinger equation is what is known as a differential equation. The ability to solve it requires a knowledge of advanced calculus, so we will not write it here. It does allow us to determine how a mathematical function, known as the *wave function*, moves in both space and time given the forces (potential energy) acting. Due to the nature of the equation, all possible allowed energies of the system automatically come out of the solution. It should be noted that this last sentence implies that while there can be a range of energies, not any energy is necessarily allowed. That is, the Schrodinger equation leads to the quantization of energy.

One of the first systems to which Schrodinger applied his equation was the hydrogen atom, where the force between the proton and the electron is simply Coulomb's law for the attraction between oppositely charged particles. Dramatically the energies predicted from his equation were quantized and, in fact, were the same as Bohr had found in his model of the atom. In other words, his equation predicted the correct atomic spectra, but with no added *ad hoc* assumptions. In fact, the quantization of the angular momentum was also predicted from the equation. With the success of the Schrodinger equation for the hydrogen atom and other systems, physicists had a way of calculating the wave behavior of any system. We now know how the wave function moves in space and time, but what exactly is it (i.e., what is waving)?

It was Max Born who gave the meaning of the wave function that is still accepted today. Born suggested who the wave function was related to the *probability* of finding the system in a given state. While this last sentence might sound rather innocent at first, it has been one of the most controversial and far-reaching ideas in physics. It certainly deserves further discussion. One technical point, to be accurate: it is the square of the wave function that is the probability. For us, this does not really change the nature of the discussion.

Why is this probability idea so strange? To begin with, all of the equations of classical physics lead to definite predictions. For instance, if we have a ball subject to a given force, Newton's second law allows us to calculate exactly where that ball will be at any chosen time. We get a unique, single solution for the location of the ball. In general, if we start out with a given initial condition, with specific forces, afterwards, we expect to find the system in a unique state. But for quantum systems, we find two very different outcomes. First, the system can end up in one of possibly many allowed states, and all we can predict is the probability of finding the system in each one.

To be sure we understand this, let us consider doing the same experiment many times. Each time we start out with the same conditions with the same force, or forces, acting. Classically, the outcome of the many experiments will be the same each and every time. But in the quantum world, the situation is quite different. Each time we do the experiment, we may get a different outcome. We can predict the probability of each outcome, but in any individual experiment, we cannot know which of the allowed outcomes will occur.

You may be thinking to yourself at this point that, if this is the case, how can we experimentally verify that these quantum predictions are correct? The answer is that by indeed doing the same experiment many times, we can measure the probability of each outcome and test to see if the measured probabilities agree with the predicted. Also, we can see if only the predicted allowed states are observed. This, in fact, is what we do. We are able to measure these probabilities extremely well. We find excellent agreement with the quantum predictions.

13.9 DOES GOD PLAY DICE?

Einstein did not, at all, like the idea that nature is ruled by probability. He articulated this in his famous statement "God does not play dice with the universe." He felt that there were hidden variables that really controlled what was happening. We just had not discovered what they were. But once we did, we would then be able to predict exactly what the outcome of any experiment would be. The vast majority of physicists do not agree with Einstein's views on this. In fact, since his death, there have been several experiments that indicate there cannot be any hidden variables. Given these experiments, one can wonder what Einstein would believe today.

To further understand the implications of probability in quantum mechanics, let us consider a phenomenon we have considered before. In Chapter 11 we discussed the interference of light, and earlier in this chapter we discussed the evidence for de Broglie waves in observing the interference of electrons. In both cases we observe the pattern of multiple points of high and low intensity, indicative of wave interference, as shown in Figure 13.2. To make the point more dramatic, let us consider an experiment to observe interference using electrons. It should be noted that since we now know that light is composed of photons, we could as well consider the interference of the particles of light. There is really no difference whether we consider electrons or photons.

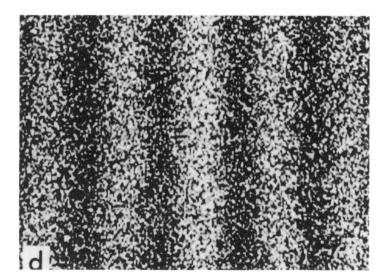

FIGURE 13.2 Interference pattern showing regions of high-intensity (light-colored) constructive interference and low-intensity (dark) destructive interference. (From Dr. Akiva Tonomura; www.hitachi.com/rd/research/em/doubleslit.html)

In the experiment we wish to consider, we will have a beam of electrons directed toward two slits separated by a distance d. The electrons pass through the slits and are observed on a photographic plate or some other device that can be used to detect the electrons. The results of the experiment are depicted in Figure 13.3.

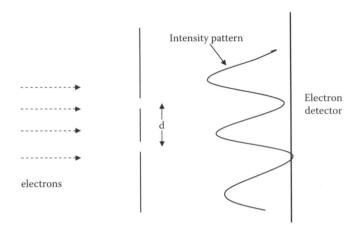

FIGURE 13.3 Electrons impinging on a screen with two slits separated by a distance d. The intensity pattern that would be observed from the result of many electrons passing through the slits is indicated.

As we discussed in Section 13.5, the Davisson Germer experiment showed that when a beam of electrons passed through the slits, an interference pattern was observed on the electron detector.

We now want to ask the following question: What is observed if, instead of shooting a beam of electrons toward the slits, we shoot only one electron toward the slits? There are only two possibilities. Either we see the same pattern as depicted in Figure 13.2 or 13.3 but much diminished in intensity, or we see a single spot on the detector. What do you think?

This experiment has been done, and the outcome may surprise you. With one electron we see one spot. As one electron at a time is shot through the slits, single spots are still observed, as shown in Figure 13.4. But with enough electrons, we start to see an interference pattern begin to form. The electrons, one at a time, tend to go to the positions of maximum intensity and do not go to the positions of minimum intensity, as depicted by the dark regions in Figure 13.2 or 13.4.

How do we understand such a result? If we solve the Schrodinger equation for the two-slit experiment, we find that the wave function gives the classical pattern of multiple areas of constructive and destructive interference. But, as we discussed above, the wave function gives us the *probability* of finding the electron. Any one electron will go to a place of high probability and be detected as a single particle. It will rarely go to a place of low probability.

These results seem to verify the probability interpretation of the meaning of the wave function, but they also lead to a rather profound question: If we see only a single

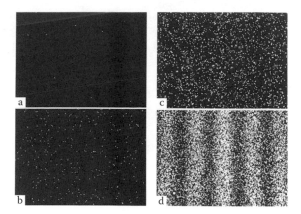

FIGURE 13.4 Buildup of interference pattern from individual electrons. (From Dr. Akiva Tonomura; www.hitachi.com/rd/research/em/doubleslit.html)

electron making a single spot on the detector, this surely implies that any one electron goes through either the upper or lower slit. After all, how can a single electron go through both slits? But, and this is very important, the shape of the interference pattern depends on the separation between the two slits. In other words, the exact location of the positions of maximum and minimum intensity depends on where the slits are. Given this information, hopefully you see what the problem is.

If a single electron goes to a position of maximum intensity, which depends on the spacing of the slits, but only goes through one slit, then how does that electron know where the other slit is? This is a profound and disturbing question.

One possible solution would be to try to detect the electron just after it passed through the slits. If you could get close enough to the slits, then you would know from which slit the electron emerged. But we have already discussed in Section 13.5 that if we want to detect something like an electron, we will have to interact with it in a way that will disturb it, as we saw in our discussion of the uncertainty principle. It can be shown that no matter how we try to detect the electron after it has passed through the slits, this act of detection will disturb the electron enough so that the interference pattern will be destroyed. If there is no interference pattern, then there is no special locations of maxima that would depend on the electron knowing where the other slit was. Over the years, there have been many very clever experiments to get around this. The bottom line is that if somehow we can gain knowledge of which slit the electron has passed through, the interference pattern disappears, so we have wiped out the evidence that the electron knew where the other slit is. In one experiment, using photons instead of electrons, information was initially obtained on which slit the photon passed through, and later on along the path that information could be erased. Without the erasure, there was no interference, but by erasing that knowledge before the photon reached the detecting screen, the interference pattern was observed (see *The Fabric of the Cosmos* by Brian Greene).

This, to be sure, sounds very strange. But it is the probabilistic quantum world we seem to be stuck with. Luckily for us, these weird effects do not appear in our macroscopic, everyday world. But remember, there are electronic devices that we probably use every day that are based on quantum mechanics. To requote Gell-Mann and Feynman, "No one really understands quantum mechanics." God does seem to play dice.

13.10 END-OF-CHAPTER GUIDE TO KEY IDEAS

- What phenomenon was a concern of the physicists at the end of the nineteenth century? Why?
- What was Planck's idea? How did it differ from what was expected classically?
- What is the photoelectric effect?
- What was Einstein's idea to explain the photoelectric effect?
- Why was Einstein's idea so controversial?
- What are the differences between the particle and wave predictions in the photoelectric effect?
- What were Bohr's rules for the atom?
- Which of Bohr's rules were new, *ad hoc* ideas?
- What is the de Broglie relation? What is its meaning?
- What was the evidence for the de Broglie hypothesis?
- Why don't we normally observe the wave nature of matter?
- What is an electron microscope? What is its advantage over an optical microscope?
- How does the de Broglie hypothesis lead to Bohr's angular momentum rule?
- What does it mean for something to be quantized?
- What is wave-particle duality?
- What is the basis for the uncertainty principle?
- What are the two uncertainty relations?
- What are the two different meanings of the uncertainty principle?
- What is the fluctuation of the vacuum?
- What is the Schrodinger equation?
- What information does the wave function give us?
- In a two-slit electron interference experiment, what is observed if a large number of electrons are incident on the slits?
- In the two-slit experiment, what is observed if a single electron is incident on the slits? Why is this so surprising?
- Does God play dice?

QUESTIONS/PROBLEMS

1. Why do concert halls full of people get hot?
2. Why was black body radiation a problem for physicists at the end of the nineteenth century?
3. What was Planck's idea to solve the black body problem? How did it differ from what was expected classically?
4. Where does the name *quantum* come from? Why is it relevant for the theory Planck proposed?

5. List all the consequences of Equation 13.1.
6. What does it mean for something to be quantized?
7. What is the photoelectric effect? List several uses for it today.
8. Why was Einstein's paper on the photoelectric effect so controversial at the time?
9. What are the differences between the wave and particle predictions in the photoelectric effect?
10. If light is a wave, why is there no threshold frequency in the photoelectric effect? If light is made of particles, why is there a threshold frequency?
11. If light is a wave, why does the time for emission in the photoelectric effect depend on the intensity? If light is made of particles, why is there no time delay?
12. What is the work function?
13. From a classical physics point of view, list two problems with the planetary model of the atom.
14. What was the primary reason that caused Bohr to propose his model of the atom?
15. Which two of Bohr's rules for the atom were based on the ideas of Planck and Einstein? Indicate which goes with which person.
16. Which of Bohr's rules were entirely new, proposed by him?
17. What is the de Broglie relation? What is its meaning?
18. Which one of Bohr's rules is consistent with the de Broglie relation?
19. What is the relevant relation between the de Broglie wavelength and the size of the electron orbit that leads to Bohr's rule 5?
20. Why can an electron microscope be used to observe objects of atomic dimensions while an optical microscope cannot be used for this purpose?
21. What is the de Broglie wavelength associated with a baseball thrown at 90 mph? (Be careful about units and make a reasonable estimate for the mass of a baseball.)
22. What is the de Broglie wavelength associated with a proton with a speed very close to the speed of light? Do not worry about any relativistic effects.
23. What is meant by wave-particle duality?
24. What is the meaning of Equation 13.6?
25. What is the basis of Equation 13.6?
26. What happens if we try to view an electron with light with a wavelength small enough so that diffraction will not limit our ability to observe the electron?
27. What happens when we try to locate an electron with light with an energy small enough so it will not disturb the electron very much?
28. What are the two meanings of Equation 13.7?
29. Describe the vacuum.
30. What is meant by fluctuations of the vacuum?
31. What evidence do we have that fluctuations of the vacuum exist?
32. Calculate the time interval for an electron-positron pair to appear and then disappear into and then out of existence.
33. What is tunneling?
34. What does the Schrodinger equation do?

35. In solving the Schrodinger equation, what properties of the system are predicted?
36. How does the solution of the Schrodinger equation differ from that obtained from an equation in classical physics?
37. What information does the Schrodinger equation give us?
38. What information does the wave function give us?
39. How do we experimentally verify predictions in quantum mechanics?
40. What did Einstein mean when he said "God does not play dice with the universe"?
41. What are hidden variables?
42. In a two-slit interference experiment, what is observed if a large number of particles are incident on the slits?
    a. Does it matter if these particles are photons or electrons? Explain your answer.
43. In a two-slit interference experiment, what is observed if a single particle is incident on the slits?
    a. Does it matter if the particle is a photon or an electron?
    b. Is this surprising? Explain your answer.
44. In a two-slit interference experiment, what happens if we try to detect which slit the particle goes through?
45. Why don't we observe quantum effects in our daily life?

14 The Standard Model of Elementary Particle Physics

14.1 INTRODUCTION

When we look at this world we live in, it appears to be a very complex place made up of millions of structures from inanimate objects like oceans, mountains, planets, and stars, to a vast variety of plants and animals, including us. Is our universe composed of an almost infinite number of different objects, or is there a basic, simple set of fundamental building blocks of matter from which everything else is made?

This is a question that goes back some 2,500 years, when Aristotle proposed that air, earth, fire, and water were the basic building blocks, and other Greek philosophers proposed the idea that atoms were the fundamental building blocks of matter. What we call the standard model is our most modern attempt to answer this age-old question. It appears to be extremely successful in that, up to now, all predictions of the theory have been experimentally verified.

We should be a bit careful in using just the words *standard model* in elementary particle physics. There is also a standard model of cosmology, which is known as the big bang theory, and there is a standard model of biology, which is known as evolution. But for the purpose of brevity, we will use *standard model* in this book to describe the one for particle physics.

The standard model goes beyond the question of matter and also attempts to answer the same question about forces. Again, when we observe our world, there seems to be a wide variety of forces: gravitational, electrical, magnetic, muscular, frictional, atomic, and nuclear, to name a few. Can we understand the large number of different forces in terms of a relatively small number of fundamental forces? The standard model says yes (remember Chapter 4)! In fact, it goes further in that it also correlates the fundamental forces with the fundamental particles of matter in that each particle "feels" only certain forces and not others.

At this point, a clarification should be made between elementary particles and fundamental particles, names that are sometimes used interchangeably and could be confused. Fundamental particles are basic building blocks that cannot be broken down into smaller entities, while another name for elementary particles could be subatomic. They are not necessarily fundamental, but some could be. For instance, protons and neutrons are elementary particles but are not fundamental since they are composed of smaller, fundamental particles. For a more detailed description of the different types of particles, see the particle physics flowchart at the end of this chapter. The various and probably unfamiliar names will be explained in the text that follows.

Ironically, an understanding of the subatomic particles, the smallest objects, is necessary to understand the largest structure we know of, the universe itself. This is because during the first fraction of a second just after the big bang (the name for the beginning of the universe), the temperatures were so immense that the only things that could exist were the fundamental particles. Any nuclei or atoms that could form would be instantly ripped apart due to the unimaginable high temperatures at the time. So, an understanding of the smallest objects is necessary to understand the evolution of the universe itself.

14.2 THE BASIC IDEAS OF THE STANDARD MODEL

A key word in understanding the standard model is *unification*. Another key word is *simplicity*. Unifying ideas should allow us to understand seemingly complex systems in terms of a few simple fundamental objects.

As indicated above, the standard model unifies both matter and forces into a small number of basic entities. As we discussed earlier, this is just what the Greeks attempted to do with their atomic model. In their picture, the atom was a small, indestructible solid ball. All matter was composed of a relatively small number of different atoms. But during the latter part of the nineteenth century, we saw that the experimental evidence indicated that the atom had to have an internal structure, so could not be fundamental. Further evidence suggested that the atom resembled a microscopic planetary system with negatively charged electrons circling a positively charged nucleus made up of protons and neutrons. So it was tempting to think that electrons, protons, and neutrons would be fundamental particles. Three basic particles sounded very nice; it was a simple system. But surprisingly, as physicists started exploring the nucleus with newly invented particle accelerators, they discovered more and more elementary particles. By about 1950, there was a zoo of over 100 known particles. A new unifying idea was needed, and it came in the form of suggesting a new set of fundamental particles called *quarks*. But before we can understand these new particles, we will need to first consider forces.

14.3 THE UNIFICATION OF FORCES

In Chapters 4 to 6 we have discussed the unification of forces and the idea of the field to allow us to understand how forces can be transmitted between objects that are separated by a distance. The field concept has proven to be very powerful and is used today for most purposes, such as radio, television, or microwave transmission. But in our most modern theory, each force field is associated with its own force particle. In this picture, a particular force is created between two matter particles when the appropriate force particle is exchanged between them. Figure 14.1 may help in visualizing how this comes about. It shows two ice skaters approaching each other from opposite directions. As they pass, one skater throws a heavy ball to the other. Each one will have his or her path altered due to the exchange of the ball. If the ball was invisible and you were just watching the two skaters, it

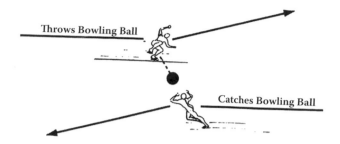

FIGURE 14.1   Two ice skaters exchanging a bowling ball making it appear that there is a force of repulsion between the skaters.

FIGURE 14.2   (**See color insert following page 102.**) In our most modern view of forces, two electrons exchange a photon (γ), the force carrier associated with the electromagnetic force.

would look like there was some force between them causing them to slightly repel each other.

The diagram in Figure 14.2 shows how we now understand the electromagnetic force between two charged objects, in this case two electrons. The electromagnetic force particle is called the photon. It is the particle Einstein first mentioned in his explanation of the photoelectric effect. It is also the particle associated with electromagnetic waves, which includes radio waves, microwaves, x-rays, and light. In order to have a force between two electrically charged objects, a photon must be exchanged. Such diagrams are known as Feynman diagrams, named after their inventor, Richard Feynman, who was a Nobel Prize winner and one of the great physicists of the twentieth century. These diagrams are very useful in help-ing us visualize how elementary particles interact via the different fundamental forces.

14.4   BOSONS: THE PARTICLES ASSOCIATED WITH FORCES

Since we have four fundamental forces today, we have four fundamental force parti-cles, which are given the generic name bosons. While these differ in important ways, as shown below, they have one property in common. They all have integer values of a basic unit of a quantity called spin. Elementary particles can spin around an axis

TABLE 14.1

The Fundamental Interactions. The column labeled "Intrinsic Strength" shows how strong the forces are relative to the strong nuclear force. The mass unit of GeV for the weak bosons is a commonly used unit for elementary particles. In this unit, the proton has a mass of about 1 GeV.

Force	Intrinsic Strength	Boson Name	Symbol	Mass	Charge	Spin
Strong	1.0	Gluon	g	0	0	1
Electromagnetic	1/137	Photon	γ	0	0	1
Weak	10^{-9}	Weak Boson	$W^{\pm}, Z^0$	~ 100 GeV	$\pm e, 0$	1
Gravitational	10^{-38}	Graviton	G	0	0	2

just as the earth spins around its axis. Spin is just angular momentum (see previous chapter). The smallest, basic unit is Planck's constant, h. All elementary particles have either integer (0, 1, 2, etc.) or half integer (1/2, 3/2, etc.) units of spin (h). Those particles with integer spin are known as bosons, while those with half-integer values are called fermions. We will come back to this when we talk about matter particles. Table 14.1 summarizes the four fundamental interactions and the properties of the corresponding force particles.

There are several things we should notice about this table. First is that gravity is so much weaker than any of the other forces. We have already touched on this in Chapter 5. But even with it being so weak, if there is enough mass in a relatively small space, it can dominate all the other forces. This is what happens in a black hole.

The particle associated with the strong nuclear force has been named the *gluon* because it acts like a strong glue holding the nucleus together. While it has no mass and no charge like the photon, it behaves very differently. It acts over very small, subnuclear distances, it is not sensitive to electric charge, as the photon is, and has a strength over 100 times that of the photon.

On the other hand, the particle responsible for the weak nuclear force is different in several ways. It comes in three varieties: two charged, W^+ and W^-, and a neutral particle, Z^0. Also, all three have a mass of about 100 GeV (100 times the mass of the proton). The fact that they have such large masses is part of the explanation as to why the weak nuclear force is so weak. In addition, the W boson can be considered a decay-inducing particle. It is responsible for nuclear β-decay, as was discussed in Chapter 10. The Z^0 being neutral acts more like the photon in producing a force between particles. We will discuss this below.

14.5 ELECTROWEAK UNIFICATION

We briefly mentioned in Chapter 4 that we have unified the electromagnetic and weak nuclear forces. We are now in a position to try to understand this. The

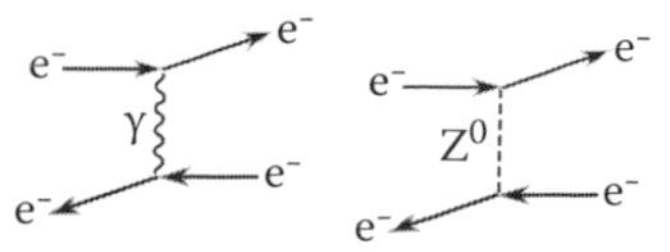

FIGURE 14.3 (**See color insert following page 102.**) Feynman diagrams depicting the force between two electrons due to the exchange of a photon (electromagnetic force) and the exchange of a Z^0. When the kinetic energies of the electrons are much greater than the mass-energy of the Z^0 (100 GeV), the force due to either exchange becomes the same. For energies below 100 GeV, the two forces do not behave the same.

information in Table 14.1 will help us. First, let us refresh our memories about the mass units used in the table. According to Einstein's famous equation $E = mc^2$, mass and energy are equivalent. But the important use of the energy unit is to allow us to compare the mass-energy of a particle to its kinetic energy of motion. When the kinetic energies of two particles with different masses are much greater than their mass-energies, we can completely ignore the difference in masses as far as energy is concerned. Of course, if there are other properties that are different, e.g., electric charge, then that could make a difference as to how the particles would interact.

Now, let us look at Table 14.1. We see that the Z^0 boson and the photon (γ) have the same electric charge (0), but the former has a mass of 100 GeV, while the latter is massless. This is essentially the basic difference between the two. At low energies (i.e., energies comparable to or less than 100 GeV), this makes a huge difference in how the photon or Z^0 would affect the interaction between, let us say, two electrons. But at energies that were available right after the big bang, much greater than 100 GeV, the Z^0 and photon would cause the same interaction between any two charged particles, as depicted in Figure 14.3. Thus, as far as the electrons in the figure are concerned, at high energies they would feel the same force from the exchange of a Z^0 or a photon.

14.6 THE UNIFICATION OF MATTER

As stated earlier, the question of the fundamental structure of matter dates back to the early Greeks. Figure 14.4 depicts a more modern view. We see that such small things as viruses are made up of molecules, which are made up of atoms, which are made up of electrons and a nucleus. The nucleus itself is made up of protons and neutrons, which are made up of entities called quarks. This picture could keep going, with the quarks being made up of something even smaller. But all the evidence we have today indicates that the quarks are indeed fundamental. This is an essential part of the standard model.

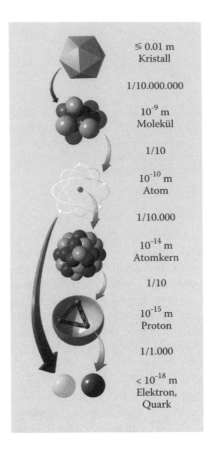

≤ 0.01 m
Kristall

1/10.000.000

10^{-9} m
Molekül

1/10

10^{-10} m
Atom

1/10.000

10^{-14} m
Atomkern

1/10

10^{-15} m
Proton

1/1.000

$< 10^{-18}$ m
Elektron,
Quark

FIGURE 14.4 **(See color insert following page 102**.) The ultimate makeup of matter. (From Desy; www.zms.desy.de/e548/e550/e6943/e83/imageobject148/kristall_quark_hr_ger.jpg)

14.6.1 Two Classes of Matter Particles

Note that in the description above, electrons were not indicated as being composed of quarks. They are not. In fact, they, and other particles like them, form a separate class of fundamental particles known as *leptons*, which is derived from the Greek *leptos*, meaning "fine" or "small." They interact via the weak nuclear force. Since some of them are charged, like the electron, they also interact via the electromagnetic force. Figure 14.5 shows a chart of the quarks and leptons. Let us consider the similarities and differences between these two types.

14.6.2 Similarities

Both are fermions (they all have a spin of ½ unit); there are six types (called flavors) of each, and the flavors are grouped into three pairs of families. For the quarks, the up and down quark flavors are in the first family, charm and strange form the second family, and top and bottom the third. For the leptons, the electron and the electron neutrino form the first family, the muon and its neutrino the second, and the tau and its neutrino the third.

Matter Particles

Charge

Leptons $\begin{pmatrix} e \\ \nu_e \end{pmatrix} \begin{pmatrix} \mu \\ \nu_\mu \end{pmatrix} \begin{pmatrix} \tau \\ \nu_\tau \end{pmatrix}$ $\begin{matrix} -1 \\ \\ 0 \end{matrix}$

Quarks $\begin{pmatrix} u \\ d \end{pmatrix} \begin{pmatrix} c \\ s \end{pmatrix} \begin{pmatrix} t \\ b \end{pmatrix}$ $\begin{matrix} +2/3 \\ \\ -1/3 \end{matrix}$

+ Anti-particles

FIGURE 14.5   The twelve matter particles: six leptons and six quarks.

Another similarity is that both quarks and leptons have antiparticles associated with them. Antiparticles are, in one sense, mirror images of their particle partners. For instance, the antielectron has the identical mass as the electron, but all other properties, such as electric charge, are opposite. So the antielectron has a positive electric charge and is, in fact, called a positron. We will refer to antiparticles briefly below.

14.6.3   DIFFERENCES

One of the basic differences is that quarks are never seen alone, but always are bound together in groups of two or three to form observable particles such as protons. On the other hand, leptons, e.g., the electron, are directly observable.

Electric charge provides another difference. The leptons, in the upper row in Figure 14.5, e, μ, τ, all have one unit of negative electric charge (the magnitude of the charge of the electron or proton is defined as the basic unit), while the neutrinos are uncharged. On the other hand, the upper quarks, u, c, t, all have $+2/3$ of the unit charge, while the lower quarks, d, s, b, all have $-1/3$ of the unit charge.

Finally, they also differ in their basic interaction. As stated above, the leptons interact through the weak nuclear force. The quarks interact through the strong nuclear force. Those leptons that are charged also interact through the electromagnetic force. Quarks also interact through the weak nuclear force in that one quark can decay into another flavor via the exchange of a charged weak boson, W^+ or W^-. Most also have mass, but we can neglect the gravitational force since it is so minute compared to the others.

14.6.4   MORE ABOUT QUARKS

One of the important differences pointed out above is that individual quarks are never observed alone. This is because the strong nuclear force is so strong that it prevents quarks from separating too far and getting free. Quarks only combine in two classes of observable particles.

In one pattern, the quarks combine in groups of three, forming particles known as *baryons*, where the name comes from the Greek word *barus* meaning "heavy." Both protons and neutrons are baryons. The proton is made of two *up quarks* and one *down quark* (note that the charges of the three quarks, $2/3 + 2/3 - 1/3$, add up to one unit of charge). The neutron, on the other hand, is made up of two down quarks and one up quark (its total charge $2/3 - 1/3 - 1/3$ is zero). Figure 14.6 shows a representation of the proton, with the quarks being held together by the exchange of gluons.

The other combination pattern is where a quark and an *antiquark* combine to form particles known as *mesons*, from the Greek word for "middle." These particles are not as familiar to us as protons and neutrons, and are only observed when produced in particle accelerators or cosmic rays. Their masses tend to be less than that of baryons.

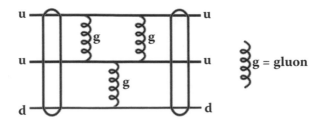

FIGURE 14.6 **(See color insert following page 102.)** A proton is made up of three quarks: two up quarks and one down quark. The quarks are held together by the exchange of the strong force carriers, the gluons.

14.6.5 MORE ABOUT LEPTONS

As stated above, leptons interact via the weak nuclear force. The e, μ, and τ leptons, having electric charge, also interact via the electromagnetic force. Since that force is so much greater than the weak nuclear force, it dominates their interactions. On the other hand, the neutrinos, all having no charge, interact only through weak nuclear force. Since this force is so weak, neutrinos hardly interact at all. This has important consequences for us. Electron neutrinos are copiously produced in nuclear reactions that fuel all burning stars, including, of course, our sun. Since they hardly interact, they escape from the sun, depositing very little, if any, of their energy back into the sun. This, in fact, is the main mechanism for star cooling. If the neutrinos interacted more, additional energy would be deposited in the star, causing it to burn faster. If this were true for our sun, it would have burned up all its fuel by now and would be a dead star. In other words, we could no longer exist, at least here on earth. So, the existence of weakly interacting neutrinos is very important to us.

14.7 A MYSTERY

Just about all of the observed, normal matter in the universe is composed of atoms, which in turn are made up of electrons, protons, and neutrons. The latter two are composed of up and down quarks only. In addition, we have just discussed the importance of electron neutrinos. It would appear that the universe only requires four fundamental particles: up and down quarks, the electron, and the electron neutrino. So, why does nature need four other quarks and four other leptons? At this time, we do not know the answer. It is one of the mysteries that particle physicists are trying to unravel.

14.8 PARTICLE FLOWCHART

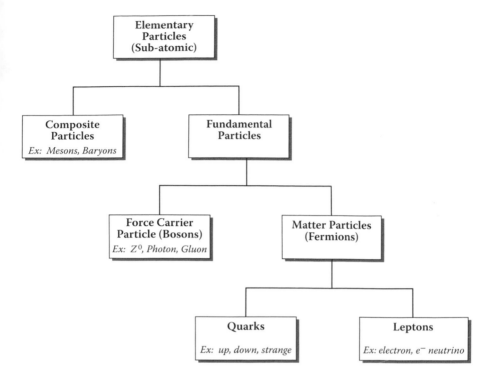

FIGURE 14.7 Particle physics flowchart showing the distinction between elementary particles and fundamental particles.

14.9 END-OF-CHAPTER GUIDE TO KEY IDEAS

- What was one of the earliest attempts in understanding the basic makeup of matter?
- What is the standard model?
- What is the difference between a fundamental particle and an elementary particle?
- What is the link between fundamental particles and the early evolution of the universe?
- What is an important idea (word) that is necessary in understanding the standard model?
- What is our modern understanding as to how forces arise?
- What are bosons?
- What are the names and properties of the bosons associated with the fundamental forces?
- How do we understand electroweak unification?
- How many classes of matter particles are there? What are they called?
- How many leptons are there? What are their names and properties?

- How many quarks are there? What are their names and properties?
- What are the similarities between quarks and leptons? What are the differences?
- Why is the neutrino so important to our existence? Which one?
- What is the purpose of the second and third lepton and quark families?

QUESTIONS/PROBLEMS

1. Describe the standard model.
2. What is the difference between fundamental particles and elementary particles?
3. Why is an understanding of fundamental particles necessary for an understanding of the evolution of the early universe?
4. Name other standard models.
5. What is meant by *unification* as it pertains to the standard model? What quantities are unified in the standard model?
6. Where have we seen the idea of unification earlier in this book?
7. Name at least five elementary particles. Which are fundamental?
8. Give the classical description as to how forces arise.
9. Give the modern description as to how forces arise.
10. What are bosons? Give all the properties of each separate boson.
11. List the four particles associated with the four fundamental forces. What one property do all these particles have in common?
12. How much stronger is the electromagnetic force than the gravitational force?
13. How much stronger is the electromagnetic force than the weak nuclear force?
14. How much weaker is the gravitational force than the weak nuclear force?
15. List three things that are different about the weak boson compared to the other bosons.
16. List two things that the electromagnetic and gravitational forces have in common.
17. What one thing do the weak and strong nuclear forces have in common that is different from the electromagnetic and gravitational forces.
18. Explain eletroweak unification.
19. Why, at energies well above 100 GeV, do the electromagnetic force and weak force act the same, but at energies below this, they do not?
20. Draw the Feynman diagram for the electromagnetic force. Explain in your own words what this means to you.
21. Draw the Feynman diagrams for both the electromagnetic force and the weak nuclear force (for Z^0 only). Explain in your own words what diagrams have to do with electroweak unification.
22. How do we know that quarks are fundamental and not composed of something even smaller?
23. How many classes of matter particles are there? How are they similar and how are they different?
24. What is the difference between families and flavors?

25. List all the fundamental forces that can affect the electron? Do the same for the electron neutrino?
26. List all the fundamental forces that can affect quarks?
27. What are the quarks that compose the proton? The neutron?
28. As stated in the text, mesons are particles that are made up of a quark and an antiquark. Mesons have either neutral or +1 or −1 unit charge.
    a. Ignoring the first sentence above, would it be possible for a meson to be made up of a u and a d quark? Explain.
    b. What would be the charge of a meson that was made up of a u quark and a d antiquark?
    c. What would be the charge of a meson that was made up of a u quark and a c antiquark?
29. What would be the charge of a baryon that was made up of three u quarks? Of three d quarks?
30. Why is the neutrino so important to our existence? Which one?
31. Which fundamental particles make up all of the normal, observed matter in the universe?
32. How many total lepton and quark families are there? Which ones seem necessary for our existence? Explain.

15 Cosmology

15.1 INTRODUCTION

Have you ever looked out your window on a clear, star-lit night and wondered: What is out there and how did it get there? I know that I did when I was a teenager. It was that intense curiosity about what was in the heavens and how it all began that caused me to become a scientist. If you have never experienced that clear, star-lit night, you have missed one of the most beautiful and profound aspects of nature. The darker your surroundings, the better, since it will allow you to see many more stars and experience the true beauty of the heavens.

When I looked out at the universe that star-lit night, our knowledge of the universe at that time was rather limited. We did know that our sun was part of a galaxy, which is sometimes called the Milky Way. We knew that our galaxy was one of millions or billions of other galaxies, and our own galaxy was pretty typical. We also knew that the universe was expanding in the sense that the galaxies were all moving away from each other. In addition, the farther any galaxy was away from us, the faster it was moving with respect to us.

Cosmology, the history of the evolution of the universe, was not a very exact science at all. In fact, there were two very different cosmological theories at that time. One was the big bang, which theorized that the universe began at some instant in time via an unimaginable explosion. The other, the steady state theory, was just the opposite. It posited that there was no beginning or end of the universe. On average, it looked the same at any time at any place. Because of the known expansion of the universe, there was one other interesting aspect of this theory. Matter had to be continually created to keep the universe always looking the same. So, this theory was also known as continual creation. Besides the experimental evidence of the expansion, there was very little other observational data available. In fact, experimental cosmology was essentially a misnomer.

Today, this has changed dramatically. There is now a standard model of cosmology, and the experiments to verify this model have attained accuracies of better than 1 part per million. The rest of this chapter will be devoted to understanding our present view of the universe and how this has come about. We will start by discussing one of the things that was known when I was a teenager: the expansion of the universe.

15.2 THE EXPANSION OF THE UNIVERSE

Before about 1920, astronomers thought that the universe was made up of the Milky Way galaxy, a collection of randomly moving stars a few hundred light years across. They also believed that the universe was static. In other words, while the stars were moving, there was no coherent motion of either a net contraction or expansion. This view was dramatically changed by one man, Edwin Hubble, for whom the Hubble telescope is named. In just a few years he showed that there were other galaxies at distances much farther away than the size of the Milky Way. Even more surprising, he showed that these galaxies were moving away from us, and the farther they were away, the faster they were receding: the universe was expanding. In fact, his measurements showed that the universe was expanding in a very regular way. Namely, the speed of a galaxy was directly proportional to its distance from us. In other words, if you make a graph of speed versus distance, it will be a straight line. This relationship is known as Hubble's law. It is considered one of the great discoveries of the twentieth century. In fact, in formulating his general theory of relativity, Einstein had built in the fact that he believed the universe had to be static. After Hubble's discovery, Einstein stated that it was the greatest blunder he had ever made.

Hopefully Hubble's discovery that the more distant a star, the faster it is moving away has evoked two questions in your mind:

1. How do you measure the speed of a star that is millions or billions of light years from us?
2. How do you measure the distance of stars? Especially very distant stars. This may sound easy, but think about measuring the distance of a very small object that is moving directly away from you. In fact, the distance measurement has proved to be one of the most challenging in modern cosmology.

Let us discuss question 1 first.

15.2.1 Measuring Speeds Using the Doppler Effect

If you have ever experienced the change of pitch from the siren of a police car or ambulance as it first approaches and then goes away from you, you have experienced the Doppler effect. It is a wave phenomenon that occurs when there is relative motion between the source of the wave and an observer. If the source of a wave is moving toward the observer, the observer will perceive the wave frequency to be higher (or the wavelength shorter) than that emitted by the source. If the source is moving away, the observer will perceive a lower frequency (longer wavelength) than that emitted from the source. For a mechanical wave (one that requires a medium), the shift of frequency is different, depending on whether the source is moving relative to the medium or the observer is moving. But for light, since there is no medium, there cannot be such a difference. The Doppler shift equations in relativity are given by the equations

$$f = f_0 \sqrt{(1 + \beta)/(1 - \beta)} \qquad (15.1)$$

$$f = f_0 \sqrt{(1 - \beta)/(1 + \beta)} \qquad (15.2)$$

where β, as in Chapter 12, is the ratio of the relative speed to the speed of light. Equation 15.1 is for the case when source, observer, or both are moving toward each other, and Equation 15.2 is when source, observer, or both are moving away from each other. f_0 is the proper frequency emitted by the source, and f is the Doppler shifted frequency measured by the observer. In the first case, the light is shifted to higher frequency. We say the light is *blue shifted* since blue light is at the high end of the visible light spectrum. In the second case, we say the light is *red shifted* since red light is at the lower end of the visible spectrum. It should be made clear that in the former case, it is not that the light is blue, but rather that the shift is to higher frequency. Similarly, in the latter case, the light is not red, but the shift is to lower frequency. In fact, we still refer to the light as being blue or red shifted even if the light is not in the visible part of the electromagnetic spectrum.

We will not be using these equations in any detailed manner, but they do tell us several important things. First, if we know the value of f_0 and measure f, then we know the relative speed, β, between source and observer. Even if we do not know f_0 but the source is moving away from us at one time and toward us at another time, we know when f is lower, the source is moving away, and when f is higher, it is moving toward us. This might be the case, for example, in observing the light from a star that is in orbit about some other astronomical object.

Also, these equations tell us that the Doppler shift depends only on the relative speed, β, and not on any other quantity, such as distance, for example. Thus, if we are looking at the light from distant stars and can measure the Doppler shift, then we know the speed of the stars independent of the distance from us. In fact, we can tell nothing about the distance using the Doppler effect.

> Hopefully this discussion of the use of the Doppler effect to measure speed has evoked a question in your mind.

If a shift in frequency is measured, how do we know what it is shifted from? If we measure f in Equation 15.2, then what is f_0? After all, in order to calculate β, we have to know both f and f_0.

The answer goes back to what we discussed in both Chapter 2 and Chapter 7, namely, the symmetry of space translation. When we look at the light from a distant star, we are looking at the light emitted from the *atoms* on that star. Because we believe in the symmetry of space translation, we believe the atoms of stars anywhere in the universe behave the same as the atoms here on earth. Their atomic spectra are the same, and thus we know f_0. While this is very straightforward, the answer to question 2 is more complicated.

15.2.2 MEASURING DISTANCES

We are talking about measuring distances of millions or even billions of light years from us. A light year is about 6×10^{12} miles, and the closest star to us is about 4 light years away.

For nearby stars, we can measure distances by the use of geometry based on the earth's motion in its orbit. For more distant stars, we have to use entirely different techniques.

15.2.2.1 Nearby Stars

For stars close enough to us, we can use simple geometry based on the earth's motion around the sun. This is known as the method of geometric parallax. Figure 15.1 shows the earth's position, relative to the sun, 6 months apart. The arrowed lines represent

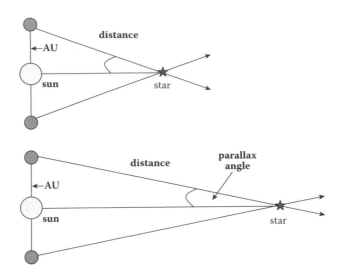

FIGURE 15.1 **(See color insert following page 102.)** Two examples of parallax measurements: one for a relatively nearby star and one for a star farther away. The parallax angle is smaller the farther the star.

the directions we would have to train our telescope to observe the same star 6 months apart. (This corresponds to a distance of the diameter of the earth's orbit.) The lengths of the lines represent the average distance to the star, which can be easily calculated by dividing the diameter of the earth's orbit by the subtended parallax angle shown in the figure. As can also be seen, as the distance to a star increases, its parallax angle decreases. To use common convention, the parallax angle is defined as half of the angle shown in the figure. If the base of the triangle used is the average radius of the earth's orbit (known as the astronomical unit, AU), then the distance to the star is given by

$$d = AU/\alpha \qquad (15.3)$$

The most common distance unit in astronomy is based on this relation. The parsec (pc) is defined as the distance to a star whose parallax angle is 1 arc-second (arcsec; remember an arcsec is 1/3,600 of a degree). With this definition, Equation 15.3 becomes

$$d = 1/\alpha$$

where d is measured in parsec and α in arcsec. Also, 1 pc is 3.26 light years.

The closest star to earth is Proxima Centauri, which is at a distance of about 4 light years. Its parallax angle is 0.77 arcsec. Because parallax becomes smaller the greater the distance to a star, useful distances can only be measured for stars whose parallax is greater than the precision of the measurement. The most accurate measurements were achieved by the Hipparcos mission, which had a small telescope mounted on a satellite. It obtained a precision of a few milliarcseconds (1/1,000 arcsec), which provided distances for stars out to a few hundred parsecs. There are about 100,000 stars within this distance, which is a very small fraction of the over 1 billion stars in our own galaxy.

15.2.2.2   More Distant Stars: Standard Candles

The standard candle technique is based on the fact that the intensity from a point source of light (a distant star) decreases as $1/d^2$, where d is the distance from the star (an inverse square law). The inverse square law can be easily understood as being a consequence of the conservation of energy and the fact that the surface area of a sphere of radius d is $4\pi d^2$. Intensity is the energy per unit area, E/area. For the intensity of starlight at a distance d from the star, the intensity is proportional to E/d^2. Standard candles are classes of stars whose absolute magnitude (intrinsic brightness) is known. If we measure the apparent magnitude of a standard candle, then the ratio of absolute magnitude to apparent magnitude gives us the value of d^2, and hence we know the distance, d. What we measure here on earth is the apparent magnitude, so the problem boils down to knowing the absolute magnitude. This can only be determined for a small, limited number of stars. We will give two examples.

15.2.2.2.1   Cepheid Variables

Cepheids are a class of stars whose brightness varies with well-measured periods. They can be easily seen in nearby galaxies since their brightness is about 1,000 to 10,000 times brighter than our sun. It has been verified that the period of variation is directly proportional to the intrinsic brightness of the star. Thus, we see that Cepheids are very good standard candles. Periods vary from a few days to 100 days, with the higher the period, the brighter the star. Unfortunately, they are only useful for relatively nearby spiral galaxies for two reasons. First, they are massive stars with short lifetimes, so they are only found in galaxies where stars have been recently formed. Elliptical galaxies tend to be older, and large-scale star formation will have ceased, so there will be no Cepheids. Second, since Cepheids are individual stars, if they are too far away, they are too faint to see. We have been able to use Cepheids to calibrate distances out to about 50 million light years. This is a small fraction of the size of the universe, which is measured in billions of light years. One note of interest is that Polaris, the North Star, is a Cepheid.

15.2.2.2.2   Supernovae

Supernovae are exploding stars. The light emitted during the explosion is extremely bright and can be seen over very large distances. They are so bright that they outshine the entire galaxy in which they reside with a brightness 5 billion times that of our sun. Thus, they can be seen at distances comparable to the visible size of the universe.

There are several different types of supernovae, but one in particular, type Ia, can be used as a standard candle. It has been discovered that the intrinsic brightness depends on how the energy is emitted as a function of time from the peak output of the explosion until it decays away. Thus, if the light curve (the light output as a function of time) can be measured, the intrinsic brightness can be calculated. This is not as easy as it sounds. First, supernovae are rare. There is only about one supernovae explosion in our galaxy every 50 years. On the other hand, there are billions of galaxies. For instance, a search team from the University of California at Berkeley found twenty in about 3 years. The other problem has to do with not only observing the supernovae but having the time to be able to measure the light curve. The typical time for the decay of the light is 20 to 40 days. In order to do this in some systematic way, organized and automated searches have been initiated. This has become extremely important since they can be used to measure universal distance scales. In fact, their use has led to one of the most surprising discoveries in all of science—the acceleration of the universe. We will discuss this further a little later in the chapter.

15.3   LIGHT FROM THE BIG BANG: CMB RADIATION

One of the primary pieces of evidence for the big bang comes from an accidental discovery made in 1965. At that time, two Bell Laboratory scientists, Arno Penzias and Robert W. Wilson, were trying to develop a very sensitive receiver to detect radio signals from outside the earth. When they first turned on the receiver, they noted a noise that they thought was due to their electronics, or possibly even from pigeon droppings. After eliminating all other possibilities, they came to the conclusion that the earth was bathed in microwave radiation from outer space and very uniform in all directions. We now realize that this radiation, known as cosmic microwave background (CMB) radiation, is the remnant radiation of the big bang. At the instant after the big bang, the universe was unimaginably hot, but due to the expansion, it has cooled to an average temperature of 2.7 degrees above absolute zero (Kelvin). This is the temperature associated with the CMB radiation. This radiation originated at a time 300,000 years after the big bang, but it gives us a window that enables us to view the universe at a time something like 10^{-35} s after the big bang. We will see below how we know these times.

Before we discuss in detail the evolution of the universe, we should discuss why the discovery of CMB radiation is a death blow to the steady state theory. The expansion of the universe is certainly consistent with a big bang, but remember, the steady state theory was able to accommodate an expansion by theorizing the continual creation of matter to keep the universe still always the same. But with CMB radiation we know the universe was much hotter at an earlier time and then cooled to the temperature we see today. The universe had to evolve in time and could not always be the same, thus negating the steady state idea.

15.4   THE EVOLUTION OF THE UNIVERSE

As was said in the introduction of this chapter, we now have a well-understood standard model of how the universe evolved. The basis of this model is that the universe began in a minute region of space in an explosive event that we call the big bang.

Figure 15.2 depicts this history. What is remarkable, as the figure indicates, is that we believe we can trace that history from a minute fraction of a second after the big bang to the present day, 14 billion years later. There are a few gaps that we are not quite sure of at the earliest times. We think we know what happened but we do not yet know how it happened.

At the instant of the big bang, we believe that all of the fundamental forces were unified as one force. As the universe expanded, this one primordial force separated into the four fundamental forces we first discussed in Chapter 4. Also, at the instant of the big bang, we believe that there had to be equal amounts of matter and antimatter.

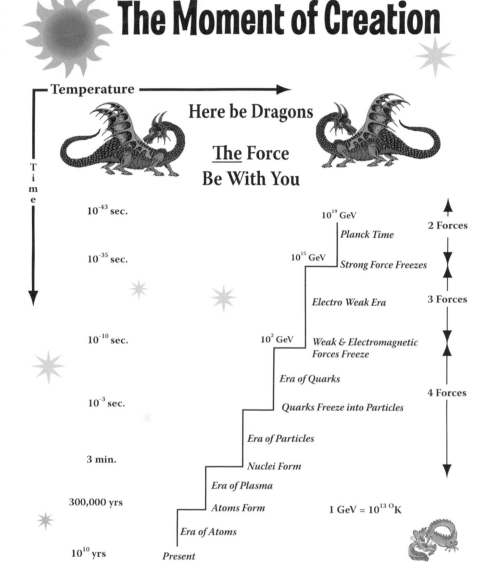

FIGURE 15.2 **(See color insert following page 102.)** The evolution of the universe from the big bang to the present. (Adapted from *Moment of Creation* by James Trefil.)

An important aspect of understanding the evolution of the universe is to realize that the big bang was an unimaginable explosion with an unimaginable temperature (of the order of 10^{32} degrees Kelvin). As time went on, as in any explosion, the universe expanded and, because of the expansion, cooled. In fact, if you remember, we discussed the separation of forces and the analogy of water cooling back in Chapter 4. It is this cooling of the universe that is responsible for the way it evolved. It allows us to pinpoint certain important epochs. We will discuss some of these epochs by taking a close look at Figure 15.2.

15.4.1 THE PLANCK TIME

The shortest time in the figure is 10^{-43} s. Why is this time significant? It is the time when we believe the primordial single force separated into the gravitational force and a second strong-electroweak force. So, right after that time there were two fundamental forces. The energy of 10^{19} GeV (1 GeV = 1,000 MeV) is known as the Planck mass (in mass-energy units). It is the mass such that the gravitational potential energy of such a mass equals its energy in mc^2 units. The time of 10^{-43} s is known as the Planck time. It is associated with the Planck mass and is the shortest time compatible with gravity and quantum mechanics. Also, there is a Planck length of 10^{-33} cm, which we believe was the size of the universe at this time.

15.4.2 THE GUT TIME

GUT stands for grand unified theory. Unfortunately, we have the name, but not a fully workable theory. Physicists believe when we have the correct theory, it will explain the separation of the strong nuclear force from the unified strong-electroweak force at the GUT time of 10^{-35} s after the big bang. As part of this theory, we believe there is a force particle associated with the unified GUT force that has been named the X boson, which has a mass of about 10^{15} GeV. For times shorter than the GUT time, and hence energies greater than the mass-energy of the X boson, the mass of the X particle would not matter and the strong force would be indistinguishable from the others. This is the same reasoning we used in Section 14.5 in discussing electroweak unification and the Z^0 mass of 100 GeV. Also, once the energy in the expanding universe dropped below the X boson mass-energy of 10^{15} GeV, the X boson could no longer be produced in collisions, and it would have disappeared just as the Z^0 would disappear at energies below 100 GeV. It should be noted that we can detect the Z^0 in accelerator experiments today because the accelerators can reach energies sufficient to produce it. We have no accelerators that can produce energies anywhere near 10^{15} GeV.

One other interesting aspect of the GUT theories is that they predict that the proton would not be absolutely stable but would decay. In fact, experiments have looked for proton decays in large vats of water. So far, no decays have been observed, which has allowed us to put a lower limit on the lifetime of a proton of 10^{33} years. This has implications for the mass of the X boson in some GUT theories. The longer the lifetime of the proton, the more massive the X boson must be.

Two other very important evolutionary events took place around the GUT time. One had to do with the disappearance of antimatter, and the other had to do with an extremely rapid expansion of what was the entire universe at that time. This expansion has been named inflation.

15.4.2.1 The Disappearance of Antimatter

In Chapter 7, we discussed the symmetry between matter and antimatter. We also discussed that symmetry had to be broken in the early universe, or otherwise, we would not exist today. As was stated in Chapter 7, all that was needed was very small asymmetry of 1 part in a billion (10^9) to explain our matter-dominated universe today. Remember, if there was a perfect symmetry between matter and antimatter, they would all have annihilated, leaving only pure electromagnetic energy and nothing else. When we look at the present universe, we can measure the amount of matter energy and electromagnetic energy and see that the matter energy is 1 part in a billion of the electromagnetic energy, not zero.

How this slight asymmetry came about, we do not know. One possibility is that antimatter particles decayed with a slightly smaller lifetime. In other words, they would have decayed at a faster rate than their matter partners. We are not sure of this, but we have observed such a phenomenon in the decay of two different particles due to the weak nuclear force. One of these, the K meson, contains a *strange quark*, and the other, the B meson, contains a *bottom quark*. While this gives us a hint of what may have happened, we know these differences in lifetimes are too small to explain the asymmetry that took place in the early universe.

15.4.2.2 Two Sticky Problems and a Solution

For many years there were two major problems that perplexed cosmologists. One was known as the horizon problem, and the other the flatness problem. In Section 15.3, it was mentioned that the CMB radiation is very uniform in all directions. This fact is the basis of the horizon problem.

This uniformity means that different regions of the universe all had to be at essentially the same temperature. But how could that be? One part of the universe would have to know about the temperature in another part that was at a distance greater than light could travel in that time. In other words, they could not be in causal contact, so supposedly could not influence each other. They were beyond each other's horizon.

The flatness problem has to do with the shape or curvature of the universe, which in turn has to do with the amount of matter there is. If the curvature of the universe is zero, we say the universe is flat. This means there is exactly the right amount of matter, known as the critical density, so the gravitational attraction of all that matter just balances out the energy of expansion due to the big bang. If there is too much matter, the gravitational attraction will win out and the universe will fall back on itself in a "big crunch." We say the shape of the universe is closed and has positive curvature. If there is not enough matter, the universe is open and will continue to expand forever.

Measurements today tell us that the density of the universe is fairly close to the critical density. But you may say "fairly close" is not good enough to tell us what the future of the universe will be. It turns out that "fairly close" is more than good enough. At 1 second(s) after the big bang, if the density of the universe differed by more than 1 part in 10^{20} from critical density, then because of the expansion of the universe, the density today would be orders of magnitude away from the critical value. So why was the universe born so flat? To physicists, this cannot be an accident. There must be a reason. This is the flatness problem.

15.4.2.3 The Solution: Inflation

The solution to both of these problems was suggested in 1979 by Alan Guth, who was not a cosmologist but, ironically, an elementary particle physicist. At first, it might seem strange that a physicist who studies the smallest, fundamental particles comes up with an idea that explains the workings of the universe. But remember, at the times we are talking about, the universe was so minute that the physics of quantum mechanics ruled.

Guth's idea was that around the GUT time, the universe underwent an extremely large, rapid acceleration—an inflation. Why did it take so long (remember our smallest quantum mechanical time is 10^{-43} s, so the GUT time is equivalent to 10^8 clicks of the universal clock at that time) and what caused this inflation? Both answers, according to Guth, were due to the fact that the universe was not born in the lowest possible energy state but in an unstable high-energy state (in quantum mechanics this is known as a *false vacuum*, where the *vacuum* is the lowest possible energy state in the system). An analogy is a coin that is balanced on its edge. If nothing disturbs it, it can stay that way. But if there is some very small perturbation, it will rapidly fall to its lower potential energy state. So, what disturbed the universe?

Just above we pointed out that the physics of quantum mechanics ruled the universe at these times. So, the disturbance was a quantum fluctuation. The universe just had to wait long enough for the fluctuation to be large enough to knock it off its edge. But once that happened, it rapidly expanded. How rapidly? Well it was an unimaginable rate. In about 10^{-34} s the universe's size doubled nearly 100 times. To put it another way, after the expansion, the universe was 10^{25} times larger than before the inflation. It was about a centimeter in size.

If you use our old friend $d = vt$, or $v = d/t$, you will see that the speed of this expansion in 10^{-34} s is much greater than the speed of light. Does this violate relativity? The answer is no. What expanded was space itself. Relativity forbids objects with mass to travel at the speed of light, or more. Space has no mass, so inflation does not violate relativity.

How does this inflationary idea answer our two problems? For the horizon problem, remember that just before the inflation, the universe was exceedingly small. Any one part of the universe was very close to any other part. All parts of the universe were in communication and would be at the same temperature. Inflation, being symmetric, maintained the uniformity of temperature. Also, inflation automatically predicts that our universe is flat. That is because the inflation was so great that any part of space is stretched flat. Consider being on the surface of a balloon that is being blown up (inflated) to a tremendous size. Any region near you would appear flat because of the great expansion of the surface. This, by the way, implies that our visible universe may be just a very small part of the entire universe.

We have spent a good deal of space talking about the GUT time. Three very important, major events happened during this time: the separation of forces into the strong force and electroweak force, the disappearance of antimatter, and inflation. It is tempting to think that these are all associated with each other. Perhaps they are. Unfortunately, at this stage in our understanding, we do not know.

15.4.3 THE ELECTROWEAK TIME

We have already mentioned this in Chapter 14, when we discussed how the electromagnetic and weak nuclear forces are unified. At a time of about 10^{-10} s after the big bang, the universe had cooled to an energy of 100 Gev (equivalent to a temperature of 10^{15} degrees Kelvin). As we discussed in Section 14.5, for energies above this, the Z^0 boson could be easily produced and the weak nuclear and electromagnetic forces were unified, acting as a single force. For energies below 100 GeV, the Z^0 could no longer be produced and the electromagnetic and weak nuclear forces separated, and there were the four fundamental forces we see today.

15.4.4 THE FORMATION OF PARTICLES

Right after the electroweak time, there was still enough energy such that quarks could remain free. But at about 10^{-3} s, the universe had cooled enough so that quarks could form into the elementary particles, such as protons and neutrons.

15.4.5 THE FORMATION OF NUCLEI

Before 3 min, if protons or neutrons formed into nuclei, the kinetic energies of collisions were great enough to prevent these nuclei from staying together. After 3 min this was no longer the case, so nuclei were able to form and stay together. Only the lightest nuclei, such a deuterium, helium, and lithium, formed at this time. Our understanding of nuclear physics allows us to predict how much of these elements were formed at this time. That prediction checks out very well with the abundance of these elements we see today.

15.4.6 THE FORMATION OF ATOMS

Compared to the short times we have been discussing so far, the universe had to wait a relatively long time of 300,000 years for atoms to form. As we discussed above, for the formation of particles and nuclei, the universe had to cool sufficiently so that collisions were not energetic enough to knock those entities into their constituent parts. The energy to free an electron from the proton in the hydrogen atom is 13.6 eV. On the other hand, the energy required to separate the proton and neutron in deuterium is 2.2 MeV, a factor of almost 200,000. The universe had to cool by this amount so collisions would not knock the atom apart. This took about 300,000 years.

Since we are made of atoms, this is certainly a very important time for us. But there is also another significant aspect associated with the formation of atoms. Before this time, protons and electrons existed as a plasma of separate particles. Being charged, they easily absorbed electromagnetic waves, i.e., light. Once atoms formed, the universe was made up of neutral particles and was transparent to light. The light that was able to travel freely at that time is the CMB radiation we see today, that we discussed in Section 15.3.

The CMB gives us an accurate picture of the nature of the temperature and density of the matter in the universe at that time. We also mentioned the CMB radiation in our discussion of the horizon problem, where we noted that this radiation is extremely uniform in all directions. What do we mean by "extremely uniform"? If we look on a scale of variation of the order of 1 part in 10,000, the radiation is indeed uniform (i.e., it shows no variation). But if we look on a finer scale of 1 part in 100,000, we see definite fluctuations. Without these seemingly small fluctuations, there would be no stars or galaxies or us.

15.4.7 THE FORMATION OF STARS AND GALAXIES

Stars and galaxies are lumpy structures. If, initially, the matter in the universe was truly uniform, then there would be no preferred direction for gravity to cause matter to clump. But the fluctuations in the CMB radiation tell us that there was indeed an initial lumpiness in matter 300,000 years after the big bang.

Where did this lumpiness come from? It was the quantum fluctuations we mentioned in our discussion of inflation. The relative size of these fluctuations was maintained throughout the inflationary period and the normal expansion that followed it. Thus, the CMB radiation that originated at 300,000 years after the big bang has imprinted on it the nature of the universe at the time of 10^{-35} s after the big bang. Amazingly, it gives us a window that allows us to peer into the universe a minute fraction of a second after its creation.

So, these slight asymmetries in the density of matter allowed gravity to form clumps. These clumps, having more matter in them than their surroundings, attracted even more matter. Eventually enough matter clumped together to form stars, and these stars then formed into galaxies. We believe the first stars appeared several tens of millions of years after the big bang. The first galaxies appeared a few hundred million years later. To bring our story closer to home, our solar system was formed about 5 billion years ago, 9 billion years after the big bang.

15.5 DARK MATTER

It sounds like our story should be over and we should be very proud of our ability to understand this amazing universe in which we live. But nature seems to have surprises for us and has taught us not to get too cocky. At times, it appears that the more we know, the more we realize we do not know.

As one example, let us go back to our understanding of star and galaxy formation. We can measure the amount of matter in the universe by simply measuring the number of stars (luminous matter), which is quite easy since we can see them. We cannot see planets, but a planet's mass is minute compared to that of a star. The problem is that knowing the amount of matter and given our understanding of gravity, we find there is not enough matter to allow stars and galaxies to form. There would have to exist a good deal more matter than we have observed.

We now know that indeed there is another form of matter that is not luminous. We can only detect it by the gravitational effect it has on luminous matter. It does not interact electromagnetically, so it neither absorbs nor emits light. It is known as dark matter, for obvious reasons. We have detected it by measuring its effect in several different phenomena: by the rotation rate of stars in the outer regions of galaxies, by the relative motion of galaxies that are nearby each other, and by what is known as gravitational lensing, where gravity causes light to bend. In the first two cases, the motion of the stars or galaxies depends on the amount of gravity caused by matter. In the case of lensing, the amount of bending depends on the amount of gravitational matter. The information from all of these sources agrees in that there is about five times as much dark matter in the universe as luminous matter. This is enough to explain star and galaxy formation.

So, what is this dark matter? The answer at this time is that we do not know. We have some possible theories, but they all involve new, esoteric particles. There are several experiments that are presently under way that are designed to detect dark matter. But at this time, we have no evidence as to what it is.

15.6   DARK ENERGY

With the discovery of dark matter it would appear that we have a very good understanding of our universe and its evolution. But as mentioned above, nature seems to have surprises waiting for us.

In discussing the flatness problem above, it was mentioned that if the universe is flat, which inflation predicts and measurements confirm, then there must be just the right amount of matter, which we call the critical density. While dark matter seems to solve the galaxy formation problem, the total amount of dark and normal matter adds up to only about 30% of the critical density. Where is the other 70%? In 1999, a set of observations suggested an answer. This answer is even more mysterious than dark matter.

In Section 15.2, we mentioned Ia supernova as one type of a standard candle. It was pointed out that since supernovae are so bright, they allow us to look at the largest distances possible, which also means the earliest times possible. What was observed using the supernova data was that the rate of expansion of the universe at earlier times was less than that at later times. In other words, the universal expansion is accelerating. This was considered one of the most surprising results in all of science.

If the universe is accelerating, this suggests a repulsive or antigravity type of force. It turns out that Einstein's theory of general relativity can accommodate just such a force. In fact, when Einstein first formulated his theory, he thought the universe was static. In order for his equations to be consistent with a static universe, he had to add in a term that he called the cosmological constant. As was mentioned earlier in this chapter, when Hubble discovered that the universe was expanding, Einstein did away with the cosmological constant and called it the greatest blunder he had ever made.

With the discovery of an accelerating universe, the cosmological constant has been resurrected. It acts like an energy, but what the source of this energy is we do not know. Hence, we have named it dark energy, analogous to the dark matter whose origin we do not know either.

At this point, you may be wondering what this dark energy has to do with the missing 70% of the matter needed to make the universe flat. Well, in both special and general relativity, mass and energy are equivalent ($E = mc^2$). In fact, precision measurements on the CMB radiation have confirmed that indeed the universe is flat, and that 70% of the needed mass-energy is in the form of dark energy.

So, here we are at the end of our story about the evolution of the universe. I hope I have convinced you that we have a very good and consistent theory of the history of this universe we live in. The only problem is that, as of the time of this writing, we do not really understand the makeup of 95% of it. As a physicist, this is truly exciting.

15.7   END-OF-CHAPTER GUIDE TO KEY IDEAS

- What is the evidence for the expansion of the universe?
- What is the Doppler effect?
- What is a standard candle?
- What is the CMB radiation?
- What is the Planck time? Planck length? Planck mass?
- What is the GUT time?
- What significant events occurred around the GUT time?
- What is inflation? What problems did it solve and how?
- What is the electroweak time? What happened at that time?
- What was the process that made it impossible for particles, nuclei, and atoms to form before their "time"?
- How much time after the big bang did atoms form?
- What significant event happened once atoms formed?
- About how long in time was it that stars and galaxies formed after the big bang?
- About how long ago did the solar system form?
- Why are there fluctuations in the CMB radiation?
- What problem did the existence of dark matter solve?
- What is the ratio of the amount of dark matter to normal matter?
- What are two important properties of dark matter?
- How much of the universe is made up of dark and normal matter?
- What is the rest of the universe made of?
- Even though we do not know what it is, what is one property of dark energy?

QUESTIONS/PROBLEMS

1. What is cosmology?
2. Name two different cosmological theories. Describe both.
3. What is the evidence for the expansion of the universe?
4. How did the steady state theory take into account the expansion of the universe?
5. Describe fully the method we use to measure the speed of distant stars. Just do not say we use _____, but describe any subtleties that are required in the explanation.
6. Describe the parallax method for measuring the distance to stars. What are the limitations of this method?
7. What is a standard candle? What is the basic principle as to why a standard candle is useful.
8. List two standard candles and briefly describe what they are.
9. Why is a Cepheid variable a useful standard candle?

10. What are the limitations on the use of a Cepheid variable?
11. What is a supernova?
12. What type of supernovae can be used as a standard candle? Explain why.
13. Discuss the challenges of using supernovae as standard candles.
14. What is CMB radiation?
15. How does the presence of CMB radiation show that the steady state theory is wrong?
16. What is the temperature of the universe at this time?
17. Why does the universe cool down in its evolution?
18. How much time after the big bang was the CMB radiation able to propagate freely throughout the universe? Why this time and not another?
19. What is significant about the Planck time?
20. What is significant about the GUT time? List three separate events that occurred at or near the GUT time.
21. What would the universe look like today if there were no asymmetry between matter and antimatter?
22. How do we know that the asymmetry between matter and antimatter in the early universe was about 1 part in a billion?
23. What is one possible explanation how the matter-antimatter asymmetry came about?
24. What is the X boson? What are its properties? Why is it significant?
25. Explain how the X boson could be responsible for the separation of the strong nuclear force from the electroweak force.
26. What is the significance of the fact that the lifetime of the proton has been measured to be greater than 10^{33} years?
27. Explain in your own words the horizon problem and how inflation solves it.
28. Explain in your own words the flatness problem and how inflation solves it.
29. Explain in your own words the inflationary idea.
30. Explain the mechanism that allowed the weak and electromagnetic forces to separate.
31. The formation of particles, nuclei, and atoms all have a common explanation. What is it?
32. When we refer to the energy of the universe, what do we mean by this?
33. What was the energy of the universe when nuclei formed?
34. What was the energy of the universe when atoms formed?
35. What is the full significance of the time about 300,000 years after the big bang?
36. Explain how the formation of atoms allowed light to freely travel, while before this it could not.
37. What is the significance of the fluctuations in CMB radiation?
38. From where did the fluctuations in the CMB radiation originate?
39. What do the fluctuations in the CMB radiation have to do with star and galaxy formation?
40. What is the significance of dark matter?
41. Since we cannot see dark matter, how do we know it exists?
42. Name three different pieces of evidence for the existence of dark matter.

43. Explain why we need dark matter to be able to understand star and galaxy formation.
44. What fraction of the universe is ordinary (made of atoms) matter? What fraction of the universe is dark matter?
45. What is the significance of the type Ia supernovae observations?
46. How do the type Ia supernovae observations lead to the necessity of dark energy?
47. Who was the first one to suggest that there might be something like dark energy? What was the reason for suggesting it?
48. What is one important property of dark energy?
49. Since we really do not know at this point what dark energy and dark matter are, could they be the same thing? Explain your answer.

Epilogue

Now that you have finished the course based on this text, I hope you do have a greater appreciation of this universe in which you live. I also hope you have a greater understanding of what physics is and the methodology used to make the many and diverse discoveries mentioned in this book (i.e., what physicists do and how they do it).

In the introduction, I presented a quote by Warren Weaver. I reproduce it below with the hope that you will think about it carefully and it will be even more meaningful to you now:

> Pure science is not technology, is not gadgetry, it is not some mysterious cult, it is not a great mechanical monster. Science is an adventure of the human spirit: it is an essentially artistic enterprise stimulated largely by disciplined imagination, and based largely on faith in the reasonableness, order and beauty of the universe of which man is a part.

I hope you realize that physics is not just a collection of disconnected topics, but that there are unifying ideas and principles connecting much of what we have discussed. Indeed, mother nature is quite efficient in that there are a relatively small number of laws that govern this seemingly vast and complex universe.

Finally, in addition to having a better understanding of the concepts associated with physics, I hope you have obtained a much greater appreciation of the role of questions and have become a much better questioner in general. Curiosity is certainly the engine that drives science, but I hope your experience here will have enhanced your curiosity in all aspects of life. The best statement I know about curiosity is due to Albert Einstein. As my final thought, I end this book with his inspirational words.

> The most beautiful and most profound emotion we can experience is the sensation of the mystical. It is the sower of all true science. He to whom this emotion is a stranger, who can no longer wonder and stand rapt in awe, is as good as dead. To know what is impenetrable to us really exists, manifesting itself as the highest wisdom and the most radiant beauty which our dull faculties can comprehend only in their most primitive forms—this knowledge, this feeling is at the center of true religiousness.

SUGGESTED FURTHER READINGS

The four books listed below are all texts currently being used in courses aimed at nonscience majors. In one way or another they cover the same material as this text, but their approach, especially as related to the ideas of unification and questioning, are quite different. But you may find them useful in helping your understanding of a particular concept just because the author has chosen a different way of approaching that concept.

Conceptual Physics by Paul Hewitt, published by Addison Wesley
Physics: A World View by Larry Kirkpatrick and Greg Francis, published by Brooks/Cole.
Physics: Concepts and Connections by Art Hobson, published by Prentice Hall.
Physics Matters by James Trefil and Robert Hazen, published by Wiley.

The following are not textbooks, but if this course has heightened your interest in physics, you might find one or more of these both fun and illuminating.

The Fabric of the Cosmos by Brian Greene, published by Alfred Knopf.
Genius, The Life and Science of Richard Feynman by James Gleich, published by Pantheon Books.
The Inflationary Universe by Alan Guth, published by Addison-Wesley.
The Pleasure of Finding Things Out by Richard Feynman, published by Perseus Books.
The Whole Shebang by Timothy Ferris, published by Simon and Shuster.

Index